AF413547

COLD ATMOSPHERIC PLASMAS

Their Use in Biology and Medicine

COLD ATMOSPHERIC PLASMAS

Their Use in Biology and Medicine

Spencer Kuo

New York University, USA

NEW JERSEY · LONDON · SINGAPORE · BEIJING · SHANGHAI · HONG KONG · TAIPEI · CHENNAI · TOKYO

Published by

World Scientific Publishing Co. Pte. Ltd.

5 Toh Tuck Link, Singapore 596224

USA office: 27 Warren Street, Suite 401-402, Hackensack, NJ 07601

UK office: 57 Shelton Street, Covent Garden, London WC2H 9HE

British Library Cataloguing-in-Publication Data
A catalogue record for this book is available from the British Library.

COLD ATMOSPHERIC PLASMAS
Their Use in Biology and Medicine

ISBN 978-981-3279-84-1

For any available supplementary material, please visit
https://www.worldscientific.com/worldscibooks/10.1142/11264#t=suppl

Desk Editor: Christopher Teo

Typeset by Stallion Press
Email: enquiries@stallionpress.com

Dedication

to my parents,
Wen-Hsiu and U-Lan Kuo

and to Sophia Ping Kuo,
my wife, constant companion and best friend.

Preface

Plasma, the fourth state of matter, contains electrically charged particles as well as neutral particles: they include electrons, ions, and atoms/molecules. Thus plasmas generated by electric discharges can convert electric energy into the kinetic energy of electrons, enabling them to excite and dissociate the composed gases in the plasma, and to generate chemically reactive species, which then promote biological interactions. Low temperature plasmas generated in an open environment become practical for biomedical applications, such as bio-disinfection/sterilization, bio-decontamination, wound blood coagulation, and wound healing. This is an emerging area, opening the door to research into exploring biological effects of cold atmospheric plasmas (CAPs) as well as the development of innovative devices that offer promise for healthcare and for emergency treatment of human beings.

This textbook was prepared to help students become familiar with CAP generators, as well as to understand their operating principles, and the characterizations of these devices. The intention is to enable interested students to apply the devices, or even to design their own, in order to conduct studies and research in a fascinating area called plasma medicine. The capacity of CAPs in different biomedical application areas is demonstrated by in-vitro and in-vivo experimental tests, and is explained based on the experimental results. The text material is a compilation of research results from the past decade. Experimental approaches and analyses provide students with adequate information to help them plan their own experimental assays and tests in their current studies and in their future work. The selection of topics and the focus given to each provide a way for a lecturer to cover the area and applications in a bio-medical/bio-engineering special topics course.

The first two chapters introduce several CAP generators, their generating and operating principles, as well as their potential for medical applications. Chapter 3 is devoted to the emission spectroscopy of the plasma plume. Spectrometers are introduced to measure the emission spectra of free radicals carried by the plasma. Disinfection/Sterilization which impact the risk of hospital-acquired infections (HAIs) are discussed in Chapter 4. The biocidal effects of cold atmospheric air plasmas (CAAPs) are demonstrated by tests on oral pathogens and biofilms. Together with the effectual sporicidal effect (verified in Chapter 5), new sterilization methods using CAAPs as disinfectants to eliminate hospital and public places infections would be a very helpful addition to existing methods. The application of CAP for cancer treatment is discussed briefly. It provides a starting point for an interested researcher to delve deeper into this rewarding subject.

Chapter 5 focuses on bio-decontamination. The sporicidal efficacy of CAAP as a decontaminant is examined. A simulant mimicking anthrax spores is employed in the tests; the spread plate method is introduced to evaluate the kill rate and scanning electron/atomic force microscopies are introduced to perform the morphological study of the biological reaction of the spore to the CAAP. Applications of plasma treatment for water purification and food safety are discussed briefly. Chapters 6 and 7 are devoted to the study and assessment of CAAPs' coagulation efficacy. Blood droplets were used in in-vitro tests, and pigs as the in-vivo animal models, in the assays and tests to explore the CAAP-enhanced coagulation mechanism and to evidence the bleeding control capacity. Wound healing is discussed in Chapter 8. The positive impact of CAAP is demonstrated and explained. The potential application of CAAP to the healing of chronic wounds in human beings is deliberated upon. Chapter 9 presents animal (pig) trials on the use of plasma for bleeding control. An air plasma spray (APS) is applied in these live tissue trials. The results have shown that APS can swiftly stop the hemorrhaging of serious wounds, and can downgrade tourniquet-necessary wounds.

The presentation in this book is self-contained and should be read without difficulty by those who have an adequate engineering background. The book will be useful to undergraduate seniors, graduate students and researchers in electric engineering, biomedical engineering, as well as in plasma.

I wish to express my sincere gratitude to Professor Bernard R. S. Cheo who has given me much helpful advice and kind encouragement on my scientific evolution throughout my career. I am indebted to Professor Leo Birenbaum for commenting and improving the presentation of the book. I am also grateful to Professor Heng-Chun Li for helpful discussions on the manuscript, to Alessandro Betti and, Edmund Chow for helping me to fabricate air plasma devices, to Zach Steinbock and Sean Matson for initiating several experiments, and to Benjamin Cheng, Dr. James S. C. Chao, and Dr. Paul Wang for their constant encouragement. I would like to acknowledge my collaborators, Professors Olga Tarasenko, Svetozar Popovic, and Simone Duarte, Drs. Cheng-Yen Chen, Chuan-Shun Lin, and Todd Pedersen.

Spencer Szu-Ping Kuo

List of Figures

Fig. 1.1. Cascade impact ionization; in the plot, 'e⁻', '⊙', and '⊕' represent electron, neutral particle, and positive ion.   3

Fig. 1.2. *V-I* characteristic of electric discharge.   4

Fig. 1.3. A planar image of an arc loop recorded with a CCD camera set with 20-μs exposure. The discharge path, from the central electrode (indicated by an arrow) to the outer electrode, is elongated by a vertical air flow blowing through the discharge gap between electrodes.   5

Fig. 1.4. Marks of discharges on the surface of the outer electrode, indicating that the hot spot rotates around the central electrode.   6

Fig. 2.1. (a) Schematic of VDBD device, (b) Schematic of SDBD device, and (c) a running FE-DBD device in treating skin.   21

Fig. 2.2. Schematics of a typical (a) DBD APPJ and (b) SDBD APPJ.   22

Fig. 2.3. Electric circuit model of the discharge between the electrodes shown in Fig. 2.2a.   23

Fig. 2.4.   (a) A photo showing where to insert a discharge module into a tapered microwave cavity and showing the generated microwave plasma effluent, (b) a schematic of the power supply running a module and a magnetron together, and (c) plasma plume images without airflow (left) and with airflow at a rate of 71 slm (right).   28

Fig. 2.5.   (a) Photo of a handheld air plasma spray in running, and (b) a schematic of the spray.   31

Fig. 2.6.   (a) A circuit diagram of the power supply, and (b) voltage and current functions of the discharge in one cycle.   32

Fig. P2.1.   A partially filled capacitor.   35

Fig. P2.2.   Coordinates and orientation of a charged circular ring.   35

Fig. P2.3.   (a) Circuit diagram of the power supply of a microwave oven, and (b) a schematic of a plasma generator.   37

Fig. 3.1.   (a) A video graph of APS, (b) a schematic of the measurement setup, and (c) the spatial distribution of the emission intensity at 777.4 nm of the APS; the numerical value at each contour is the $\log_{10}$ of the intensity in Rayleigh.   48

Fig. 3.2.   Spatial distribution of the emission intensity I, in Rayleigh at 777.4 nm, along the central axis (z) of the spray.   51

Fig. 3.3.   Scheme of the imaging spectroscopy.   52

Fig. 3.4.   Axial distribution of the electron excitation temperature.   53

Fig. P3.1.   Schematic of the recording process of an optic spectrometer.   55

Fig. P3.2.   Optical spectral lines and the spectral ranges of the UV bands.   56

Fig. 4.1.   Cell walls outside the cores of (a) gram-positive and (b) gram-negative bacteria.   59

List of Figures   **xiii**

Fig. 4.2. Spatial distributions of the emission intensities of the OI lines at 202.64 nm, 770.68 nm, and 777.19 nm.   67

Fig. 4.3. A photo of the experimental setup.   68

Fig. 4.4. Comparison of the growth situations between (a) the control and (b) the treated sample. The circle in (b) spots the zone of growth inhibition.   69

Fig. 4.5. A flow chart for biofilm formation tests.   71

Fig. 4.6. Changes of the composition and viability of the treated biofilms of *Streptococcus mutans* UA159 in reference to the controls; APS treatment during the formation.   73

Fig. 4.7. The architecture and surface topography of biofilms after 5 days of formation examined by environmental scanning electron microscope: (a) biofilm treated with air flow twice-daily during the formation as a control; (b) biofilm treated with the plasma effluent twice-daily during the formation. Fluorescence microscopy stained with LIVE/DEAD backlight bacterial viability: (c) mature biofilm treated with 30 s air flow as a control and (d) mature biofilm disinfected by APS for 30 s (Red: dead cells; Green: live cells).   74

Fig. 5.1. (a) Representation of the interior and exterior structure of a bacterial endospore, and (b) SEM image of a *B. cereus* spore, including its appendages ("A") and exosporium ("E").   85

Fig. 5.2. Emission of the plasma at atomic oxygen lines $^5P(J{=}3)-^5S^\circ(J{=}2)$ at ~777.19 nm, $^5P(J{=}2)-^5S^\circ(J{=}2)$ at ~777.42 nm, and $^5P(J{=}1)-^5S^\circ(J{=}2)$ at ~777.54 nm.   88

Fig. 5.3. Schematic of the experimental setup.   89

Fig. 5.4.	CFU formation after decontamination using ASMP at several distances (30, 40, 50 mm) and exposure times (2–12 seconds).	94
Fig. 5.5.	BC kill curves: data points are obtained by placing samples at three exposure distances of 30 mm (+), 40 mm (Δ), and 50 mm (*) from the cavity opening of the ASMP. Results are compared with BG kill curves reported by Herrmann *et al.* (Phys. Plasmas **6**, 2284, 1999), using APPJ (dash line) and hot gas of 175°C (dot line).	95
Fig. 5.6.	BC kill curves; data points are obtained by placing wet samples (*B. cereus* spores in water) at three exposure distances of 30 mm (•), 40 mm (○), and 50 mm (+) from the cavity opening of the ASMP.	96
Fig. 5.7.	BC kill curve: data points are obtained by placing each envelope containing a sample at an exposure distance of 40 mm from the cavity opening of the ASMP.	97
Fig. 5.8.	SEM images of *B. cereus* spores; (a) untreated and (b)–(d), exposed to the ASMP at 30 mm distance for (b) 2 s, (c) 4 s, and (d) 8 s at low (left column) and high (right column) magnification. In (a), A stands for appendages, E for exosporium.	98
Fig. 5.9.	AFM amplitude images of *B. cereus* spores at low (column A) and high (columns B) resolution; (a) untreated and (b)–(d), exposed to the ASMP at 30 mm distance for (b) 2 s, (c) 4 s, and (d) 8 s.	101
Fig. 5.10.	AFM 3D images (first row) and section analyses for the length (A), width (B), and height (C) of (a) untreated *B. cereus* spore and of the viable spores after plasma treatment for (b) 2 s, (c) 4 s, and (d) 8 s.	102

Fig. 5.11.   Images of arc loops in ASMP taken by an
              intensified CCD camera set for a 100 ms
              exposure time; arrows indicate the position of
              the central electrode.   105

Fig. 6.1.   (a) Composition of whole blood: blood plasma
              and formed elements; and (b) separation of
              whole blood (in tube 1) into two fractions
              including A) blood plasma (upper part in tube 2)
              and B) formed elements (lower part in tube 2).   114

Fig. 6.2.   Blood samples treated (a) by a heated airflow
              for 16 s; by an air plasma spray at a fixed
              exposure distance of 25 mm with three exposure
              times of (b) 8 s, (c) 12 s, and (d) 16 s; and by the
              same air plasma spray with a fixed exposure time
              of 16 s at two increased exposure distances of
              (e) 30 mm and (f) 40 mm.   117

Fig. 6.3.   Untreated controls (row 1) of whole blood (1a),
              blood plasma (1b) and formed elements (1c)
              droplet-samples; and the corresponding plasma
              treated samples (row 2).   118

Fig. 6.4.   Changes of RBC and platelet concentrations
              of plasma-treated sample with the increase of
              the exposure time from 2 to 10 s in the cases
              of two exposure distances of 25 and 40 mm.   119

Fig. 6.5.   Cartoon plots showing the clotting process in
              the APS treatment.   121

Fig. 7.1.   Photos of straight cuts (a and b) and cross cuts
              (c and d) taken (a) after more than 3 minutes
              waiting time for bleeding to stop naturally,
              (b) after bleeding stopped by a plasma treatment
              of 18 s, (c) after 4 minutes waiting time for
              bleeding to stop naturally, and (d) after the
              bleeding stopped by 13 s plasma treatment.   131

Fig. 7.2. (a) Photo showing a saphenous vein in an ear of a pig, in which a hole is to be punched and (b) photo of the punched saphenous vein after 15 s plasma treatment to block the hole from bleeding. 132

Fig. 7.3. Photos showing (a) cutting an artery and (b) letting the bleeding stop naturally; and (c) with plasma treatment and (d) stop the bleeding in half time. 133

Fig. 8.1. Recovery of an artery cut and the surrounding irritated area shown in Fig. 7.3d, after the plasma treatment. 142

Fig. 8.2. Recovery of an irritated area impinged by plasma spray for 10 s at an exposure distance of 25 mm. 143

Fig. 8.3. Comparison of the progress of the recovery of untreated (row 1) and treated (rows 2 and 3) cross cuts. 143

Fig. 9.1. (a) A portable air plasma spray and (b) a block diagram of the power supply. 157

Fig. 9.2. Composite images showing the plume of the APS (on the right) comparing with the false-color calibrated images of 777.4-nm emissions from the APS (on the left) recorded by a narrow-band-filtered CCD camera, in which the line plots in white are the emission intensities at various distances from the nozzle. The insert at the upper left corner is the emission spectrum showing a peak at 777.4 nm. 159

Fig. 9.3. APS treatment on a cut wound on the back; (a) examining the wound, and pressing on the cut artery to slow down bleeding, (b) treatment focused on the cut artery first, (c) moving APS to treat other an area of the cut, and (d) assessing the wound to confirm complete stop of bleeding. 161

Fig. 9.4. APS treatment on a curved large cut wound in the hind quarters area; (a) treating the cut arteries back and forth continuously for about 105 s, and (b) assessing the wound to confirm no more bleeding. 163

Fig. 9.5. APS treatment on an amputated leg; (a) tourniquet is applied to slowdown hemorrhage from an amputated leg, (b) APS is applied to the cut arteries back and forth continuously for more than 5 minutes, (c) after all hemorrhage is controlled, tourniquet can be removed, and (d) wound is wrapped by bandage before giving a formal treatment in the hospital. 164

List of Tables

Table 4.1 Diameter (mm) of the zone of inhibition versus the plasma treatment time. 69

Table 7.1 Bleeding control time of air plasma treatment. 134

Contents

Preface   vii

List of Figures   xi

List of Tables   xix

1 Introduction   1

   1.1 Characteristics of Atmospheric Plasma   2

   1.2 Plasma Medicine Applications   7

      1.2.1 Disinfection and sterilization   8

      1.2.2 Decontamination   10

      1.2.3 Bleeding control   12

      1.2.4 Wounds healing   13

   Problems   14

2 Atmospheric Pressure Plasma Generators   19

   2.1 Dielectric-barrier Discharges (DBDs)   20

   2.2 Atmospheric Pressure Plasma Jets (APPJs)   22

   2.3 Arc Seeded Microwave Plasma (ASMP)   26

   2.4 Air Plasma Spray (APS)   30

   Problems   34

**3 Free Radicals and Reactive Oxygen Species
in Cold Atmospheric Plasma**   38

   3.1 Biocidal Effects of Free Radicals and Reactive
Oxygen Species   38

3.2 Plasma-liquid Interaction to Generate Additional Free Radicals and Reactive Oxygen Species   41

3.3 Emission Lines of Likely Reactive Species in CAAPs   44

3.4 Emission Spectroscopy of Atomic Oxygen   47

3.5 Emission Spectroscopy of Electron Excitation Temperature   51

3.6 Impacts of Atomic Oxygen, Ozone, Nitric Oxide, and UV Radiation   53

Problems   55

4 Plasma Disinfection and Sterilization   58

4.1 Bacteria   58

4.2 Issues of Disinfection Methods   60

4.3 Plasma Disinfection Approaches   63

4.4 CAPs on Dental Issues   65

4.5 Experiments   66

4.6 Air Plasma Treatment of Oral Pathogens   66

    4.6.1 Experimental conditions   66

    4.6.2 Zone of inhibition of microorganisms by plasma treatment   68

4.7 Dental Disinfection   70

    4.7.1 Experiment preparation and procedure   70

        4.7.1.1 Plasma treatment effect on biofilm formation   71

        4.7.1.2 Plasma treatment effect on biofilm disinfection   71

    4.7.2 Experimental results   72

        4.7.2.1 On preventing biofilm formation (case A)   72

        4.7.2.2 On biofilm disinfection (case B)   72

        4.7.2.3 Microscope observation   73

4.8 Summary   75

4.9 Discussion   75

4.10 Cancer Treatment   76

Problems   78

5 Plasma Decontamination   **81**

5.1 Background   81

5.2 Spores   83

    5.2.1 Resistance to the treatments   85

5.3 Decontamination via Biological Reactions   86

5.4 Experimental Preparations   87

    5.4.1 Experimental setup   88

    5.4.2 Materials   89

    5.4.3 Sample preparations prior to and after exposure   90

       5.4.3.1 Prior to exposure   90

       5.4.3.2 Procedures of processing the samples after plasma treatment   91

5.5 Colony-forming Unit (CFU)   92

5.6 Decontamination Experiments and Results   93

    5.6.1 Dry samples on glass slide-coupons   93

    5.6.2 Wet samples   95

    5.6.3 Sample contained inside an envelope   96

5.7 Morphological Studies   96

    5.7.1 Scanning electron microscopy   97

    5.7.2 Atomic force microscopy   99

5.8 Plausible Mechanism   103

5.9 Applications in Other Emerging Areas   107

    5.9.1 Water treatment   107

    5.9.2 Food industry   108

Problems   111

6 Effect of Air Plasma on Blood Coagulation   **113**

6.1 Blood Coagulation   114

6.2 In-Vitro Tests of Air Plasma Blood Coagulation   116

6.3 Tests on Smeared Blood Samples-Cell Count Dependency   118

6.4 Mechanism of Air Plasma Blood Coagulation   120

Problems   123

7   Air Plasma Bleeding Control Study with Animal Models   **126**

7.1 Background   126

7.2 Bleeding   126

7.3 Hemostasis   127

7.4 Clotting   128

7.5 Experimental Arrangement   129

7.6 Surface Wounds   130

    7.6.1 Test 1 — straight cut   130

    7.6.2 Test 2 — cross cut   131

7.7 Vessel Wounds   131

    7.7.1 Test 3 — hole in a saphenous vein   132

    7.7.2 Test 4 — a cut to an artery   133

7.8 Discussion   134

Problems   136

8   Wound Healing   **138**

8.1 Healing Process   138

8.2 Immune System   140

8.3 Post-Operative Observation of Wound Healing After APS Plasma Treatment   141

8.4 A Plausible Mechanism   144

8.5 Discussion   145

8.6 Chronic Wounds   147

8.7 Burn Wounds   148

8.8 Wound Care   149

Problems   150

9   Advanced Bleeding Control   **153**

9.1 Background   153

9.2 Hemostasis in Combat Casualties   155

9.3 APS as an Advanced First Aid Tool   157

9.4 Animal Model Trials   160

    9.4.1 Trial 1 — a deep cut at back   161

    9.4.2 Trial 2 — a curved large cut in hind quarters area   162

9.4.3 Trial 3 — amputated leg   162

9.5 Proposed Battlefield Simulation Trials   165

    9.5.1 Proposed Trial 1 — close range shotgun wound to rear leg   165

    9.5.2 Proposed Trial 2 — close range 7.62 mm gunshot through upper rump with large exit wound through groin   166

    9.5.3 Proposed Trial 3 — dissection and laceration of femoral artery   166

    9.5.4 Proposed Trial 4 — dissection and complete severing of brachial artery   167

9.6 Discussion   167

9.7 Summary   168

Problems   169

Bibliography   172

Index   179

Chapter 1

Introduction

Plasma is referred to as the fourth state of matter, which is distinct from the solid, liquid, and gas states. It contains electrically charged particles as well as neutral particles, including electrons, ions, and atoms/molecules. Plasmas are estimated to constitute more than 99% of the visible universe. Normally, plasmas appear in gas form, but the unique features of conducting electricity and reacting collectively to the electromagnetic forces distinguish the plasma state from the gas state.

In nature, plasma temperatures and densities range from relatively cool and tenuous (like aurora) to very hot and dense (like the central core of a star). Plasmas are also generated artificially for different applications in the areas of energy production, power conversion, material processing and surface treatment, space engine, medicine, etc. Different plasmas are required in different areas of applications. In medical applications, the applicable plasma imposes criteria, which guide the design of the plasma generator. Those include 1) appearing in open air (i.e., an atmospheric pressure plasma) with large spatial extent; this feature is practically necessary in many medical applications; 2) carrying a significant amount of free radicals, such as reactive oxygen/nitrogen species (ROS/RNS); free radicals, in particular ROS, influence biochemical reactions and enhance metabolisms in therapeutic processes; and 3) low thermal temperature, so that the treatment will not damage healthy tissue and cause patient discomfort.

1.1 Characteristics of Atmospheric Plasma

At atmospheric pressure, the density of the air is about 2.35×10^{25} molecules/m^3. In the presence of high electric field, stray electrons and ions created by ionization from cosmic rays are energized to initiate cascade impact ionization, which causes electric breakdown. Cascade impact ionization requires that the electron ionization rate ν_i is larger than the electron loss rate attributing to attachment, recombination and diffusion.

In the beginning of the electric breakdown, the recombination loss can be neglected because the charge density is still low; the diffusion loss can be neglected as long as the discharge gap is much larger than the electron mean free path. However, as an electron attaches to a neutral molecule, the negatively charged molecule cannot contribute to the ionization because it is too heavy to gain enough energy from the applied electric field. Thus, the threshold field $E_{cr}(air)$ to cause electric breakdown of the air, estimated from the balance between the ionization gain and the attachment loss of the electron density, is about 3 kV/mm, where both of the ionization rate and attachment loss rate are proportional to the electron collision frequency, dominated by the electron-neutral collisions.

Electrodeless discharge can occur in microwave field. On the other hand, electrodes, which are conductors, are used as the two terminals of an electrically conducting medium, for voltage connection; it conducts current into and out of the medium, such as plasma generated by low frequency electric discharge. The electrode from which electrons emerge is called the cathode and is designated as negative; the electrode that receives electrons is called the anode and is designated as positive. The electric field distribution varies strongly with the geometry of an electrode. In air; an electrode with a sharp tip, the electric field intensity near the sharp tip is given by $E \approx q/4\pi\varepsilon_0 R^2$, where q, ε_0 and R are the electric charge accumulated on the tip of the electrode, the free space permittivity, and the radial distance from the tip, respectively; a cylindrical rod electrode, the electric field intensity, near the rod, is given by $E \approx \rho_s a/\varepsilon_0 r$, where ρ_s and r are the surface charge density on the rod and the circular radius around the rod (with $r \geq a$, the radius of the rod); two parallel planar electrodes, $E \approx V/d \approx \rho_s/\varepsilon_0$, where V is the applied voltage and d is the separation of the two electrodes. Therefore, intense fields, of the order of 10^{10} volts/m, can be generated in the neighborhood of sharp points and edges of electrodes, where field ionization occurs. However, the field intensity of a point charge drops rapidly with the distance (i.e., $\propto R^{-2}$); if the field intensity near the other electrode is dropped well below the breakdown threshold, the induced corona discharge around a sharp point may not evolve to electric breakdown. In other words, the applied voltage to the electrodes has to ensure that the field intensity near both electrodes exceeds the breakdown threshold. On the other hand, the geometry of an electrode pair can impact the electric characteristics of the discharge between the electrodes. Thus, the design of the electrodes, which makes difference in the plasma generation, is significant to the applications of the plasma generator.

After the electric breakdown with the applied electric field intensity $E > E_{cr}$ (i.e., $v_i > v_a$ where v_a is the electron attachment rate), cascade impact ionization as shown in Fig. 1.1 quickly builds up the charge density, which is governed by the rate equation

$$\frac{dN_e}{dt} = (v_i - v_a)N_e - \alpha N_e^2 \qquad (1.1)$$

where N_e is the electron density, α is the recombination coefficient, and the ion density $N_i = N_e$ is assumed. The recombination

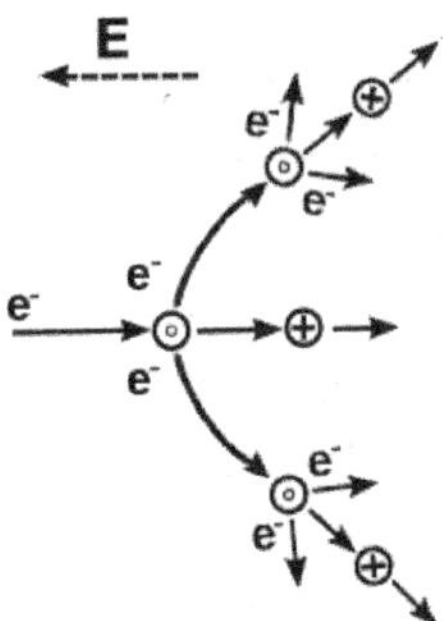

Fig. 1.1. Cascade impact ionization; in the plot, 'e⁻', '⊙', and '⊕' represent electron, neutral particle, and positive ion.

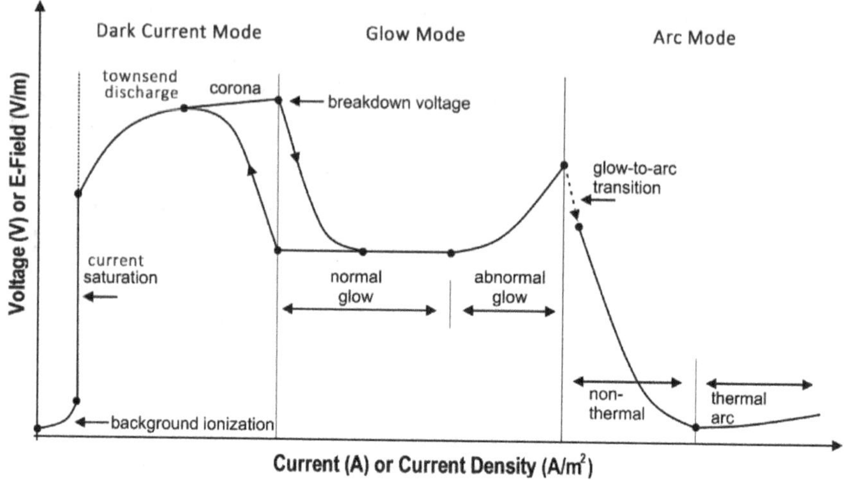

Fig. 1.2. *V-I* characteristic of electric discharge.

term on the right hand side of (1.1) needs to be retained only after the degree of ionization reaches a significant level.

As the ionization increases, it elevates the discharge current to evolve the discharge into a glow mode. As shown in Fig. 1.2, after an initial voltage drop in a transition, the discharge voltage increases with the increase of the discharge current, i.e., $dV/dI > 0$, in the glow discharge. However, if the current is not limited, the driving electric field (which has a large initial value) tends to increase the discharge current rapidly, which transits through the glow mode to establish arc discharge. In the arc mode which appears in the high current discharge region as shown in Fig. 1.2, the discharge voltage drops rapidly as the discharge current increases. It has a negative *V-I* characteristic, i.e., $dV/dI < 0$. When a current source is used, the discharge voltage will drop to a much lower level than that of the breakdown voltage. In other words, the maintenance field E of the arc discharge can be much lower than the breakdown threshold field E_{cr}.

It suggests that with the aid of a high voltage trigger pulse to initiate the electric breakdown, an atmospheric air plasma generator may be run with a relatively low voltage power supply, which has the advantages of being safe and compact.

In air, the attachment loss is much larger than the recombination loss, and the ionization rate is given to be

$$v_i = 3.83 \times 10^2 v_a [\varepsilon^{3/2} + 3.94\varepsilon^{1/2}] \exp\left(-\frac{7.546}{\varepsilon}\right) \quad (1.2)$$

where $\varepsilon = E/E_{cr}$ is the normalized electric field intensity. Because v_a in atmospheric air is large, ionization can build up rapidly after the electric breakdown to evolve the discharge to an arc mode. The arc tends to become constricted and to develop hot spots on the electrode surfaces; after the constriction, the arc becomes hot and has very small volume, which do not meet the criteria of the device design for medical applications.

To avoid arc constriction, air flow and/or magnetic field have been introduced in devices to generate plasma jet and plasma spray, where the discharge paths are elongated by the airflow, as shown in Fig. 1.3, and the discharge hot spot rotates in the magnetic field

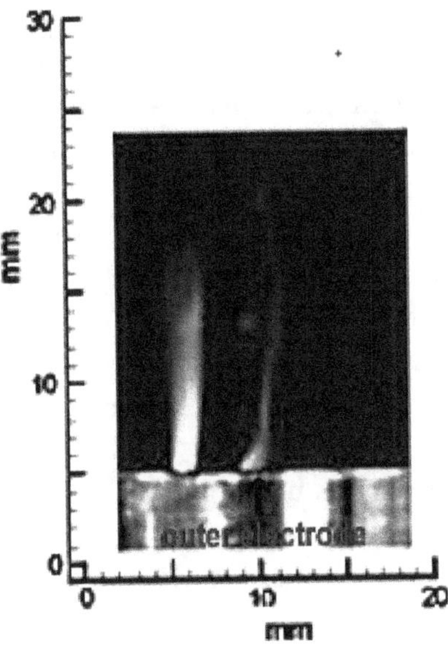

Fig. 1.3. A planar image of an arc loop recorded with a CCD camera set with 20-μs exposure. The discharge path, from the central electrode (indicated by an arrow) to the outer electrode, is elongated by a vertical air flow blowing through the discharge gap between electrodes.

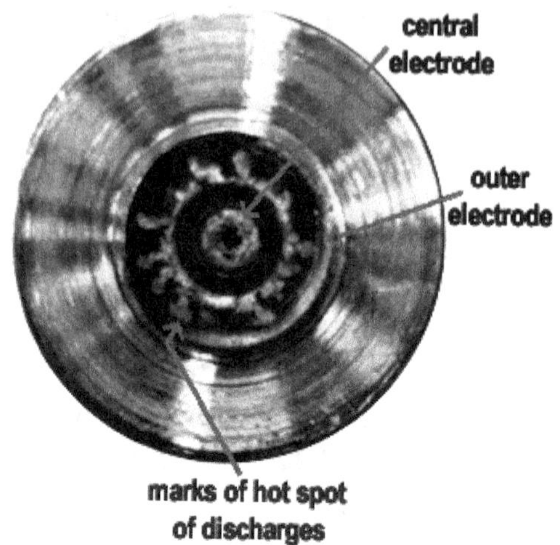

Fig. 1.4.   Marks of discharges on the surface of the outer electrode, indicating that the hot spot rotates around the central electrode.

around the central electrode, as shown in Fig. 1.4, to keep arcs diffusive. In other words, the arc column dragged by the gas flow evolves to a gliding arc mode. The non-equilibrium stage starts when the length of the gliding arc exceeds a critical value. The magnetic field rotates the gliding arc, making it diffusive and prolonging the non-equilibrium state. A periodic voltage source also works to maintain the non-equilibrium state. The advantage of running the air discharge in low voltage high current diffusive gliding arc mode is that the power supply can be compact, and an additional gas tank is not necessary, which makes it easy to design a practical portable device, as detailed in Chap. 9.

Some of the plasma jets use noble gases as the working gases to reduce the attachment rate ν_a, which in turn, reduces the breakdown threshold field E_{cr}, thus, the discharge voltage and current, and shifts the discharge to the glow mode, with the aid of injected gas flow. For example, the breakdown threshold fields of Helium and Argon gases at atmospheric pressure are $E_{cr}(\text{He}) = 0.96$ kV/mm and $E_{cr}(\text{Ar}) = 1.82$ kV/mm, respectively. Another way to limit the discharge current is to introduce dielectric barrier(s) in front of the electrode(s). In this arrangement, the discharge is generally performed by the electric field of a few kHz to tens of MHz. The discharge current is mainly the displacement current rather than the conduction current; thus the discharge has low power factor, keeping down the ohmic loss and the plasma temperature.

The function of the plasma generated by the discharges is to convert electric energy into kinetic energy of electrons, which is used for exciting and dissociating molecular oxygen and nitrogen to produce chemically reactive oxygen species (ROS) and nitrogen species (RNS), including molecular oxygen/nitrogen in metastable states and atomic oxygen/nitrogen. Atmospheric pressure air plasmas use the ambient air as the working gas, which has rich oxygen/nitrogen content and makes it easy to run the plasma generator because there is no need to prepare other gases. On the other hand, noble gases and noble gas mixed with N_2/O_2 have also been used as a way to lower the plasma temperature, but the tradeoff is lowering the intensities of generated reactive species. In any case, plasma has to be generated in non-equilibrium state; the electron kinetic energy, gained from the electric force, is better used for producing reactive species, rather than for heating the plasma effluent. The flux of the reactive species carried by the plasma is somehow proportional to the discharge current, the concentration of N_2/O_2 in the working gas, and the percentage of the energetic electrons in the electron plasma. Thus, in many medical applications, a non-thermal gliding arc discharge air plasma will be a much more effectual plasma treatment agent among the plasmas generated by different discharge schemes in various working gases.

1.2 Plasma Medicine Applications

Several cold atmospheric plasma (CAP) generators have been developed to implement plasma treatment: 1) for disinfection, sterilization, and decontamination, which inactivates or kills bacteria, fungi, viruses and spores; 2) for bleeding control, which coagulates blood swiftly; 3) for wound healing, which shortens the healing period and benefits the regeneration of the epithelization of tissue to avoid scar formation; etc. The applications in each category and the advantages of plasma treatments are briefly discussed in the following.

1.2.1 *Disinfection and sterilization*

During the course of receiving medical care, patients could be infected by a wide variety of common and unusual bacteria, mycobacteria, fungi, and viruses. However, these healthcare-associated infections (HAIs) have to be avoided; it can be done by proper disinfection of the ambient environment and thorough sterilization of medical and surgical instruments. Disinfection is to remove most of microorganisms present on surfaces and in the ambient air that can cause infection or disease; sterilization is killing or removal of all microorganisms including bacterial spores, the killing decimal log number has to meet the sterility assurance level (SAL) of 6.

The high heat wet or dry sterilization approach may not be a favorable option for all instruments because it can cause the degradation of thermolabile medical devices. Current approaches of sterilizing the instruments, such as using autoclave or dry heat oven, take time to complete the processes. Other alternative methods, such as using chemical gases or solutions, will take an even longer time. The chemicals of wet disinfectants, such as bleaches, may also be corrosive to materials such as metals, plastics, rubber, paint, leather, and skin. In addition, chemical gases and solutions may introduce side-effects, impacting environment and posing health issues. Overall, conventional methods for sterilization of medical devices take time to complete the processes; the material properties, such as molecular weight, volume, and morphology, of sensitive devices, including polymeric biomaterials, could also be altered. This can influence the physical and biological performance of the medical device, leading to material failure.

As hospitals continue to increase the number and diverse types of expensive, intricate instruments, their sterilization demands quicker turnaround times as well as reliable as top priorities.

Cold atmospheric plasmas (CAPs) are dry disinfectants; in highly energized states, CAPs contain radicals such as excited atoms and molecules. Reactive oxygen species (ROS) and atomic oxygen can destroy just about all kinds of organic contaminants by means of biochemical processes, which can inflict sufficiently high level of killing and inactivating efficacy, and do not cause apparent side effects. Thus, CAPs provide a potential way to speed up the sterilization of medical and surgical instruments as well as the clean room.

Compared to the standard disinfection and sterilization methods using pressurized hot air of 170°C, pressurized hot water vapor of 120°C, and wet chemistry, CAP sterilization method offers the following advantages:

1) Dry disinfectant, fast working and no chemical corrosive effect;
2) ultra-fine cleaning, no residues;
3) low operating temperature providing gentle sterilization treatment of medical tools and instruments;
4) no mass storage requirement and easy carrying and operating in the case of air plasma; and
5) environmental friendliness, no chemical waste products.

It has been shown that application of CAPs for hand disinfection has some advantages over conventional liquid disinfectants. In therapy of skin infections, CAPs can treat sensitive surfaces, including living tissues and open wounds; it is contact-free and painless, and does not damage healthy tissue. It has been considered to potentially treat many bacterial skin diseases such as impetigo contagiosa, folliculitis or ecthyma, fungal infections such as tinea pedis or even viral diseases.

CAPs combat two common bacteria, Pseudomonas aeruginosa and Staphylococcus aureus, which show up frequently in wound infections. Those are resistant to antibiotics because they have a protective layer called a biofilm, which is a dense aggregation of microbial cells bound together by an extracellular matrix of polysaccharide and protein. When bacterial groups aggregate in these multicellular communities, they are able to tolerate antimicrobial challenges that normally eradicate free-floating individual cells. On the other hand, CAPs are able to kill bacteria by damaging microbial surface structures and/or DNA without being harmful to human tissues. They are able to kill bacteria growing in biofilms in wounds. CAP treatment avoids the nasty side effects that drugs often bring; it destroys bacteria indiscriminately — whether it is antibiotic-resistant or not.

Another potential application is to prevent or to treat dental caries, which is among the more prevalent chronic human infectious diseases. The World Health Organization's reports describe that caries prevalence is at 60–90% in school-age children and virtually universal among adults. The etiology and pathogenesis of dental caries are generally ascribed to the colonization of tooth surfaces by multispecies biofilm containing mutans streptococci. Polysaccharides of a biofilm promote adherence and accumulation of cariogenic streptococci on the tooth surface, which contribute to the bulk and structural integrity of the matrix and enhance the pathogenic potential. On the other hand, the hydroxyl-radical, $^*OH$, and hydrogen peroxide, H_2O_2, generated via interaction of reactive oxygen species with water (H_2O), can chemically attack the outer structures of bacterial and fungal cells to cause lipid pre-oxidation, providing a strong germicidal effect.

1.2.2 *Decontamination*

Biological emergencies are those in which the microbes are released in the air causing illnesses and death. After biological contamination has been discovered, the first task is to map the contaminated area. The second step is to identify the agent's physical and biological properties.

Decontamination is then carried out by physical removal of the contaminants, or by killing or inactivation of microbial contaminants. It cleans reusable equipment, instruments, supplies, and objects in the decontamination area, and removes remaining contaminated objects from the area. The process prevents the spread of micro-organisms and other noxious contaminants that may threaten the health/live of human beings or animals, or damage the environment, and safely renders reusable items for further handling. Thus, this area and items in the area become safe, no longer posing a risk to people who come into the area and in contact with those items. In biowarfare and bioterror attack, the decontamination assurance level (DAL) is elevated to the killing decimal log number of 7.

The traditional decontamination methods for killing biological warfare agents (BWA) involve the use of "wet" solutions, which include bleaches and Decontamination Solution #2 (DS2). These hazardous chemicals need to meet special guidelines for storage, transport, and disposal during and after usage. Moreover, "green" decontaminants are preferred because these chemicals are released into the environment and may be toxic (i.e., not "green"). The decontamination time of these methods is typically 30 minutes. To reduce the exposure time, Sandia National Laboratories have used a cocktail of ordinary substances to create a type of foam, which can neutralize viral, bacterial, and nerve agents in minutes, as reported in the internal news release on March 1, 1999. A disadvantage with wet methods is that the current decontamination chemicals are corrosive to materials such as metals, plastics, rubber, paint, leather, and skin. Thus they are not suitable for use on sensitive equipment.

Alternative dry methods are being developed for many reasons, which include easily transported, fast working, no mass storage requirement, safe to personnel and inert to sensitive equipment. Ionizing radiation γ-ray is effective to decontaminate a large area in a heavily shielded fixed facility, but it is not easily implemented in the field and can be destructive to sensitive equipment.

Atmospheric pressure plasma is another alternative dry method. As pointed out in Sec. 1.2.1, this dry decontaminant is green, material friendly, ultra-fine cleaning without residues, and easy to transport and apply. More importantly, it has been demonstrated that atmospheric pressure air plasmas can kill the biological warfare agent Anthrax, and its biological simulants, effectively and thoroughly. The sporicidal mechanism is believed to be through the oxidation reduction process instigated by the atomic oxygen and ROS of the plasma effluent. Such dry decontaminant is expected to promote safety assessments in industrial and commercial applications.

1.2.3 *Bleeding control*

Bleeding, even from an external hemorrhage, may be life threatening if it is not treated swiftly. Every year thousands of trauma victims die of hemorrhage from assault, vehicle collision, workplace, and domestic accidents, as well as battlefield injuries. There is a need for developing advanced technologies to control bleeding; in particular, arterial bleeding. Gel like clot has currently been used to stop bleeding; however, it could be intrusive, in the wound healing process, to complicate the post-operation treatment.

Blood coagulation involves platelet activation and aggregation, and coagulation cascade. Oxygen free radicals promote platelet activation and aggregation; moreover, platelets involved in wound repair and blood homeostasis also release ROS to recruit additional platelets to sites of injury.

When a wound bleeds, vasoconstriction takes place, and oxidants are released in the vascular lumen to enhance platelet agglomeration at the wound location to act on blood clotting. Moreover, oxidants activate the formation of thrombin to form a fibrin clot which keeps blood coagulation in homeostasis. Thus the treatment to speed up coagulation has to reduce the blood pressure at the wound site to slow down bleeding, to clot injured blood vessels to stop bleeding, to heal the cause of bleeding and prevent complications, and to relieve symptoms.

The whole process of coagulation is a complex chain reaction, where different elements are motivated into action, conditioned from other reactions. The ability to influence these reactions speeds the whole coagulation process, which turns out also to impact the subsequent healing process.

When atomic oxygen interacts with H_2O, oxidants H_2O_2 and *OH, similar to those released in the vascular lumen, are generated. This suggests that atomic oxygen treatment could speed up blood clotting and clot formation and an atomic oxygen generator could serve the purpose.

Indeed, it has been demonstrated that an air plasma spray, which carries a significant amount of atomic oxygen, could clot uncoagulated whole blood samples in less than 20 seconds, which is much less than 30 minutes for an untreated sample to reach complete coagulation. As animal models, pigs were used in live tissue tests of air plasma bleeding control. The results show that cold air plasma can rapidly stop bleeding of critical wounds involving broken arteries, and so potentially help to save the life of an injured person under emergency situations. It also downgrades tourniquet-necessary wound, to extend the golden period of saving the remaining part below the tourniquet.

Atmospheric air plasmas applying the ambient air as the working gas has the advantages of generating ROS directly inside the working gas and making it easy to design and run a portable plasma generator because there is no need to prepare other gases.

A portable cold atmospheric air plasma (CAAP) generator could be an ideal advanced first aid for injured persons under many situations, such as on the battlefield, in open field, and in difficult moving positions. In addition, plasma bleeding control has the advantage of preventing unnecessary damage to the body, such as caused by excessive pressure or unnecessary infection caused by prolonged contact with the wound.

1.2.4 *Wounds healing*

The healing process is remarkable and complex, and it is also susceptible to interruption due to local and systemic factors, including moisture, infection, and maceration (local); and age, nutritional status, body type (systemic). When a proper method is applied for hemostasis and the right healing environment is established, the time to heal and replace devitalized tissue can be shortened.

The healing process is an interactive reaction which includes many cell metabolisms, infection prevention, which is an immediate reaction of the body responding to a wound, and growth actions. Normally, wound healing goes through four sequential overlapping phases, which are homeostasis, inflammatory, proliferating and remodeling; it takes time, which varies with the location, age, degree of wound, etc.

Because oxygen is consumed in all biological reactions and metabolisms for wound healing, hyperbaric oxygen therapy (HBOT) is a current treatment method to speed the wound healing. This mechanism uses hyperbaric oxygen to dissolve oxygen into tissue and blood, so the tissue oxygen tension around the wound is raised. The treatment is through inhalation.

Plasma treatment, on the other hand, raises the oxygen tension of the tissue and cell metabolisms at the wound site, and provides oxygen in the blood by catalytic reaction. Plasma treatment dilates local blood vessels, which speeds the immune cells to enter the damaged site. ROS carried by air plasma are dry sterilization agents, which also work for disinfection simultaneously with blood coagulation. Reactive oxygen metabolites affect thrombus formation within the vasculature and cells are aging via accumulating free radical and oxidative damage over time. In order to maintain the normal function, skin tissue is metabolized when it is aging. Because the treated area accumulates sufficient atomic oxygen, this area is aging faster than surrounding tissue. The skin tissue increases the metabolism to speed up the generation of new tissue. The new skin tissue grows under the ageing tissue, and replaces the position after aging tissue is peeled. These positive effects are expected to speed up the healing process.

Again, as mentioned in the previous section, using pigs as animal models, the healing progress of cross cut wounds was observed. The plasma treatment not only stops bleeding swiftly, but also shortens the healing time from 14 days to about 8 days. This observation suggests that CAAP may be used in dermatological treatment to speed up the healing of chronic wounds.

Problems

P1.1.　Using a simple collision model to describe the motion of an electron in a dc electric field E in the collisional environment, the momentum equation is given by

$$m\frac{dv}{dt} = -eE - vmv \qquad (P1.1)$$

where m, v, −e, and ν are the electron mass, speed, charge, and electron-neutral elastic collision frequency, respectively.

Electron is accelerated by the electric field E to gain kinetic energy, which converts to the thermal energy after each collision.

(1) A constant ν is assumed to simplify the integration, show that the gain of the thermal energy after each collision is given to be $\Delta\varepsilon_g \approx 0.2$ m $(eE/m\nu)^2$, which leads to the rate of thermal energy gain $P_G = \nu\Delta\varepsilon_g$.

Electron also loses its energy ε to the neutral particles in (elastic and inelastic) collisions. Due to the small mass ratio ($m/M_n \ll 1$, where M_n is the mass of the neutral particle) and the constraint of momentum conservation, electron only transfers a very small fraction of the energy to the neutral in each collision and the energy loss rate is given to be $P_L = \delta\nu_e\varepsilon$, where δ is the average fractional energy lost in each collision and ν_e is the effective electron collision frequency (it accounts for both elastic and inelastic collisions with the neutral particles); and $\delta\nu_e = 2\nu$ (m/M_n), i.e., $\delta = 2$ m/M_n and $\nu_e = \nu$ in the case of elastic collision. Thus a power balance equation is given by

$$\frac{d\varepsilon}{dt} = P_G - P_L = \nu\Delta\varepsilon_g - \delta\nu_e\varepsilon \qquad (P1.2)$$

Assume constant ν and $\delta\nu_e$, and set the electron initial energy $\varepsilon_0 = 0$, show that the solution of (P1.2) is $\varepsilon = (\nu/\delta\nu_e)\Delta\varepsilon_g [1 - \exp(-\delta\nu_e t)]$.

(2) At atmospheric pressure, $\delta\nu_e t \gg 1$ in a few nanoseconds, thus, $\varepsilon = (\nu/\delta\nu_e) \Delta\varepsilon_g$ is the steady state solution of (P1.2), obtained without necessary ν and $\delta\nu_e$ to be constants. To achieve gas breakdown, the free electrons in the background gas have to be energized to exceed the ionization energy ε_I of the gas, i.e., $\varepsilon > \varepsilon_I$.

Derive the breakdown threshold field $E_{cr} = E_{cr}(\nu, \delta\nu_e; \varepsilon_I)$.

(3) In atmospheric pressure air, $\varepsilon_I \sim 12.5$ eV, the ionization energy of O_2, where 1 eV $= 1.6 \times 10^{-19}$ J; in this energy range, $\delta\nu_e = 2\nu(m/M_n) \sim 3.78 \times 10^{-5}$ v and $\nu \sim 2.6 \times 10^{13}$ s^{-1}.

Evaluate the breakdown threshold field E_{cr} (V/m) in the atmospheric pressure air; $e/m = 1.76 \times 10^{11}$ C/kg can be used in the evaluation.

P1.2 Disinfection describes a process that eliminates many or all pathogenic microorganisms, except bacterial spores, on inanimate objects. On the other hand, sterilization describes a process that destroys or eliminates all forms of microbial life and is carried out in health-care facilities by physical or chemical methods. Sterilization is intended to convey an absolute meaning.

Endoscopes represent a valuable diagnostic and therapeutic tool in modern medicine. Though the incidence of infection associated with their use reportedly is very low (about 1 in 1.8 million procedures), more healthcare-associated outbreaks have been linked to contaminated endoscopes than to any other medical device

After leak testing, endoscope sterilization with a liquid chemical sterilant involves five steps: 1) Clean; 2) Disinfect; 3) Rinse; 4) Dry; and 5) Store. Explain the necessity of each step to achieve sterilization.

P1.3. THE SPAULDING CLASSIFICATION defines the minimum levels of disinfection (or sterilization) that should be employed according to the infection risk associated with a medical device. The medical equipment, devices, and facility are divided into "Critical", "Semicritical", and "Noncritical" categories for different levels of disinfection or sterilization.

Critical devices are those entering normally sterile tissue or the vascular system; those devices have to be sterile to eliminate all microorganisms including bacterial spores.

Semi-critical devices are those contacting with intact mucous membranes without ordinarily penetrating sterile tissue; those devices require high level disinfection to kill all microorganisms but allowing the possible presence of low numbers of bacterial spores.

Noncritical objects are mainly those in the facility not coming in contact with patients' mucous membranes or skin that is not intact; only vegetative bacteria, fungi and lipid viruses need to be killed.

(1) What is the killing decimal log number of sterility assurance level (SAL)?

(2) What is the required temperature to kill most human pathogens?

(3) How does heat kill microorganisms?

(4) Explain why water boiling cannot guarantee sterilization.

P1.4. Biological indicators (BIs) are microorganism controls used to monitor microbial survival probability after disinfection process. Each type of BIs has specific nominal population and D-value of the microorganism, where the "D-value", defined to be the exposition time (in minutes) required to kill 1 decimal log (i.e., 90%) of the microorganisms in the BI via heat process (121°C), is introduced to measure the microbial resistance to heat. For a BI of 2.0×10^6 at D = 1, determine the required exposition time to kill all microorganisms in the BI.

P1.5. Overkill sterilization approach is to assure safety of decontaminating big loads. A single BI provides the survival probability of microorganisms in a standard pack of load after heat process (at 121°C) of a specific exposure time, leading to the required exposition time to reduce the survival number in the pack to zero (see Prob. 1.4). When the load is equivalent to thousands of packs or more, an overkill approach by doubling the exposition time is applied. Consider a BI of *B. steorothermophilus* spores on a strip with population 1.4×10^5 and D = 1.5, calculate

(1) the exposition time for one pack of load to meet SAL,

(2) the required exposition time to eliminate all spores in one pack of load,

(3) illustratively explain why overkill approach, for a big load up to a million packs, is necessary.

P1.6. It should be aware that after a surgical procedure is over, the instruments should be pre-cleaned as soon as possible before transporting to the decontamination area. Explain why?

P1.7. A prion is an infectious protein that is misfolded. These proteins can aggregate in the brain and other neural tissue, forming amyloids that disrupt the normal tissue structure. Brain diseases associated with prions include scrapie, mad cow disease, Creutzfeldt Jacob disease (CJD) and Gertsmann-Straeussler-Scheinker (GSS) disease. Prions are resistant to most traditional methods of inactivation used for other microorganisms.

(1) Explain why prions cannot be killed, that makes difficult in decontamination; and

(2) Can prions reproduce on their own? Explain why.

P1.8. Severe blood loss, about a fifth or more of the normal amount of blood in the body, makes the heart unable to pump enough blood to the body, which causes hypovolemic shock. The most important step in treating this form of shock is to control bleeding. However, if the victim is already showing signs of shock, it is important to stabilize the victim until help arrives by taking the necessary steps: 1) Maximize oxygen delivery — completed by ensuring adequacy of ventilation, increasing oxygen saturation of the blood, and restoring blood flow; 2) Control further blood loss; and 3) Fluid resuscitation. Also, the patient's disposition should be rapidly and appropriately determined.

Describe the goal of each step of the treatment and its significance.

P1.9. Generally, adult wound healing involves fibrotic processes causing wound contraction which may lead to the formation of scar tissue, bringing varying gradations of structural imperfections, deformities and problems with flexibility. Currently, the principal goals in wound management are to achieve rapid wound closure with a functional tissue that has minimal aesthetic scarring.

It is a repair (scarring) response or a typical regeneration response (limb regeneration) in the contraction phase depends on whether the wound closure is rapid or prolonged. However, if contraction continues for too long, it can lead to disfigurement and loss of function.

Describe steps which should be taken to achieve a more perfect reconstruction of the wound area.

Chapter 2

Atmospheric Pressure Plasma Generators

Plasma generated in an open environment makes it easy for practical applications. Moreover, plasma generated in the ambient air or interacting with the ambient air can deliver abundant chemically active species, such as reactive oxygen species (ROS) and reactive nitrogen species (RNS). These reactive species are capable of destroying a broad spectrum of microorganisms as well as catalyzing coagulation cascade.

In plasma treatments for external wounds and infections, there are several criteria on the plasma generation. Those include 1) appearing in open air with large spatial extent, 2) carrying a significant amount of free radicals, such as reactive species, and 3) having low thermal temperature.

Electric discharge is a common way to produce plasma in the atmospheric pressure. As shown in Fig. 1.2, discharges in different conditions have different characteristics, which affects plasma density, temperature, distribution, uniformity, size, as well as the specifications of the power supplies and the configurations of the electrodes designed for the discharges.

Devices have been designed to apply different types of discharges in various arrangements and geometries of electrodes and in different working gases for plasma generation, which can meet the criteria. These devices generate different cold atmospheric plasmas

(CAPs), including dielectric barrier discharge (DBD) plasmas, dc/rf plasma jets, diffused arc discharge plasma sprays, and microwave plasmas, for medical applications. The main difference between DBD and other types of CAP devices is that in DBD the plasma treatment is in-situ, i.e., plasma agents directly contact with the treated tissue, while other CAPs treat samples remotely by the plasma effluent. In the following, descriptions of these devices are presented.

2.1 Dielectric-barrier Discharges (DBDs)

Dielectric-barrier discharges (DBDs) are usually driven by high voltage AC fields with frequencies in the kHz to MHz range. A conventional DBD device comprises two planar electrodes with at least one of them covered with a sheet of dielectric insulator such as alumina or quartz; the electrodes are separated by a small gap where discharge occurs.

When only one electrode is covered by a dielectric insulator, as shown in Fig. 2.1a, the applied high voltage field instigates volume dielectric-barrier discharges (VDBDs), which generate plasma in the gap between electrodes. On the other hand, when both electrodes are covered by a dielectric insulator as shown in Fig. 2.1b, only micro-discharges are generated on the dielectric surface; plasma generated by these surface dielectric-barrier discharges (SDBDs) appears on the surface of a dielectric layer in each half cycle, and can be more uniform and dense than that generated via VDBD.

A dielectric-barrier blocks the current flow; charges accumulate on the surface of the dielectric sheet to induce an electric field counterbalancing the applied field in the gap. The discharge is then quenched. However, on the next half cycle of the AC field, the induced field of the surface charges reinforces the applied field, causing a new discharge to be ignited. In other words, an AC field can maintain dielectric-barrier discharges, and the duty cycle of the discharge increases with the frequency of the AC field.

Plasmas generated by VDBDs and SDBDs are confined in bounded regions, where the space is too narrow into which to place live tissues to perform treatment. For medical application of DBD devices, the tissue/skin itself then serves as a floating-electrode, as

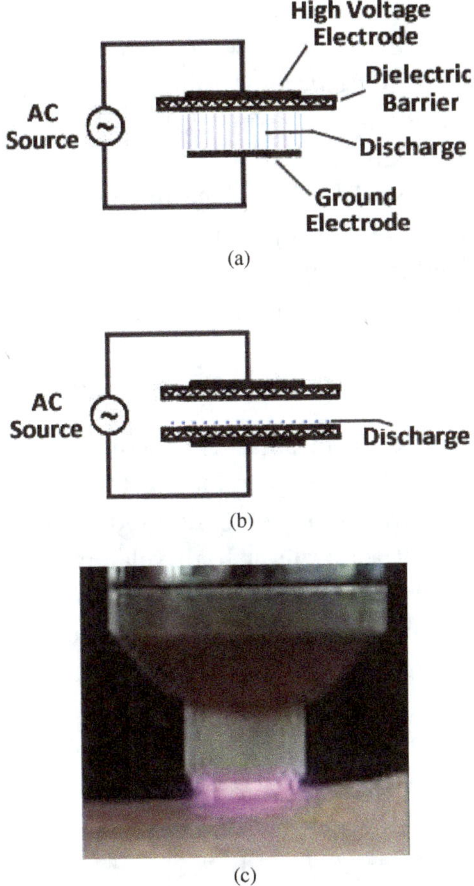

Fig. 2.1. (a) Schematic of VDBD device, (b) Schematic of SDBD device, and (c) a running FE-DBD device in treating skin.

shown in Fig. 2.1c. Such a floating-electrode dielectric barrier discharge (FE-DBD) device is essentially using tissue/skin to replace the dielectric-shielded ground electrode of a SDBD device; both adopt non-conducting ground electrode. However, FE-DBD setup becomes flexible in placing it on the surface of treating tissue/skin, which makes it adaptable in treating skin diseases and wounds. An AC voltage source of a few kV (up to about 10 kV) at a few tens of kHz (20 to 35 kHz) is applied to the dielectric barrier electrode; the produced electric field charges the nearby (a few mm) floating electrode to promote corona discharge on the surface of the floating electrode. In medical treatment, the generated micro-discharges appear on the surface of the tissue/skin, and the discharge current penetrates into the skin.

The treatment usually takes place in the room air. However, a protocol to prepare the treating tissue/skin as a proper ground electrode may be needed in the operation of the device, in order to achieve the desired effect of the treatment on the patient.

2.2 Atmospheric Pressure Plasma Jets (APPJs)

An atmospheric pressure plasma jet consists of a gaseous mixture flowing at a high rate between two electrodes, which discharge by a radio frequency electric field. The generated plasma carries reactive species to stream out a nozzle at high velocity as a jet. There is a large variety in APPJ designs used in experiments. Many APPJs use a dielectric to limit current, just like in a DBD, but not all do.

Illustrated in Fig. 2.2a is a schematic of a typical DBD APPJ, which is composed of a cylindrical capillary dielectric tube

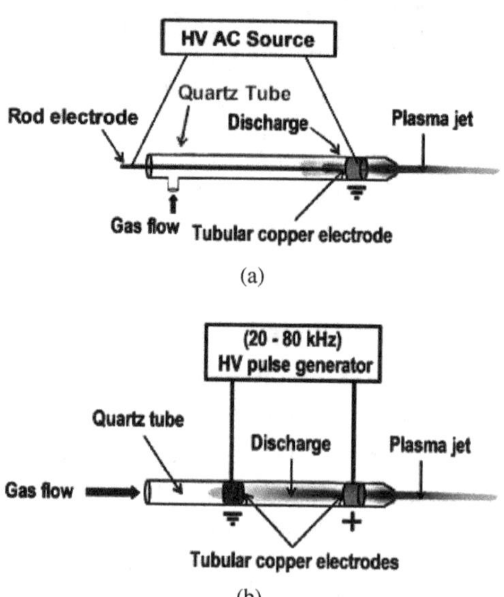

Fig. 2.2.   Schematics of a typical (a) DBD APPJ and (b) SDBD APPJ.

(quartz), and a pair of electrodes consisting of a central pin-type electrode mounted in the capillary (usually with inner diameter of 1.6 mm and outer diameter of 4 mm) and an external tubular ground electrode assembled around the tube to set up DBD inside the tube.

The gas (with a flow rate of a few slm) used is usually helium or argon, sometimes with a small amount (<5%) of O_2, H_2O or N_2 mixed in to increase the production of chemically reactive atoms and molecules. Noble gas has smaller breakdown threshold field E_{cr} and lower electron attachment rate v_a, which makes it easy to produce a low temperature stable electric discharge.

A high voltage (HV) AC power supply of various frequencies (e.g., a few kilovolts at a few MHz up to 15 MHz) is used to drive the discharge. An intense plasma is generated inside the glass tube between the two concentric electrodes, and is pushed out as a jet by the gas flow into open space through a nozzle.

The discharge current may be modelled as the internal leakage current of a capacitor. Let C be the capacitance of the concentric electrodes and R be the internal resistance of the capacitor after the discharge occurs. The voltage $V(t) = V_0 \cos \omega t$, applied to this series connected RC circuit as shown in Fig. 2.3, drives a current $I(t) = CdV_C(t)/dt$, where the capacitor voltage V_C is determined by the equation set up by the Kirchhoff's voltage law (KVL) to be

$$Vc + RC \frac{dVc}{dt} = V(t) = V_0 \cos \omega t \tag{2.1}$$

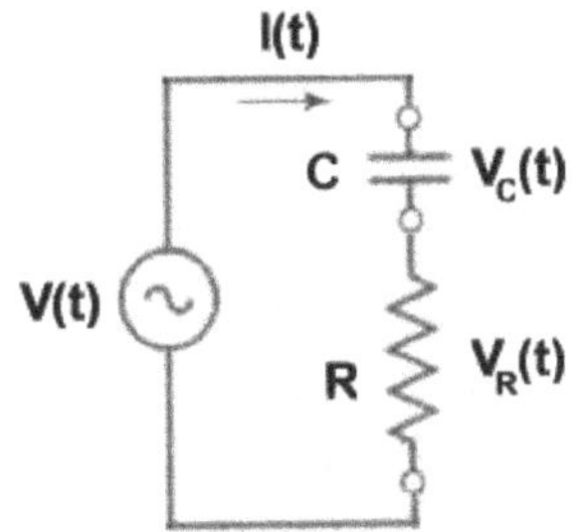

Fig. 2.3. Electric circuit model of the discharge between the electrodes shown in Fig. 2.2a.

This equation (2.1) is solved to obtain

$$V_C(t) = \left\{\frac{V_0}{1+(\omega RC)^2}\right\}[\cos\omega t + \omega RC\sin\omega t] \qquad (2.2)$$

With the aid of (2.2), the discharge current is derived to be

$$I(t) = -\left\{\frac{\omega C}{1+(\omega RC)^2}\right\}V_0[\sin\omega t - \omega RC\cos\omega t] \qquad (2.3)$$

where a constant R is assumed to simplify the analysis. Thus the time average electric power delivered to plasma (represented by R), generated by the discharge, is given by

$$P_{av} = \langle I^2 R\rangle = \frac{1}{2}R\frac{(\omega C)^2 V_0^2}{1+(\omega RC)^2} \qquad (2.4)$$

where $\langle \ldots \rangle$ represents time average. With the aid of a *V-I* characteristic of the discharge similar to that presented in Fig. 1.2, the resistance R can be determined self-consistently. Normally, the power factor $\omega RC/[1 + (\omega RC)^2]^{1/2}$ is small, the discharge power increases with the square of the frequency.

In practice, the high frequency power supply uses transformers in the components of electrical circuit, the reactance X_L of the secondary of the transformers is positive and increases with the frequency, i.e., $X_L = \omega L$, where L is the inductance of the secondary. On the other hand, the reactance X_C of the DBD electrode capacitance in Fig. 2.3 is negative and is inversely proportional to the frequency, $X_C = -1/\omega C$. Therefore, a properly designed variable frequency power supply can provide a resonant operation by adjusting the operating frequency to achieve $X_L + X_C = 0$. At this frequency $\omega_0 = 1/\sqrt{LC}$, the power supply is driving a resistive load, so that the power delivered to the plasma is maximized.

Plasma needle has similar components in the setup, except that the central rod electrode has a needle tip outside the quartz tube to generate corona/glow discharges (depending on the power of operation) and the external tubular ground electrode is located downstream a distance away from the tip of the central electrode. Plasma generated by the corona/glow discharges is blown by the injected gas flow into a needle shape.

The design of the device is simplified by replacing the central rod electrode with another external tubular copper electrode assembled around the tube, as shown in Fig. 2.2b. In this installation, both electrodes are covered by the dielectric, similar to those used in SDBD. The gap between electrodes is typically a few centimeters with Helium as the working gas, which has a breakdown threshold field $E_{cr}(\text{He}) = 0.96$ kV/mm. HV pulses at 1–10 kV and at a frequency of tens of kHz (20–80 kHz) are applied to the electrodes. The electric field outside the tube is too low to cause air breakdown. The electric field of the pulse induces polarization charges on the inner surface of the quartz tube near the tubular electrodes, where corona discharges are initiated to seed the subsequent glow discharge between the two electrodes. The glow discharge inside the tube generates ionization waves that begin inside the tube and are pushed out by a gas flow through a nozzle into open space. Electrons follow ions to form ambipolar motion, where ions are accelerated by the pulsed electric field. Plasma is in the form of a series of ionization pulses or "plasma bullets", which deliver reactive species, charged species, and photons to treat the sample. The plasma bullets have a velocity much greater than the gas flow velocity. The length of each bullet depends on the operational parameters (e.g. applied voltage, gas flow rate) and the nozzle.

A plasma pencil is designed by replacing the external tubular electrodes with dielectric shielded ring electrodes inserted inside the quartz tube. Specifically, dielectric disks of the same diameter as the inner diameter (2.5 cm) of the quartz tube and having a center hole each for streaming flow, hold and cover a copper ring each to form dielectric electrodes. The HV dc pulses ionize the injected gas flow to generate plasma which spurts out the tube in a pencil-shape plasma jet.

Because the working gas is helium or argon, sometimes with a small amount (<5%) of O_2, H_2O or N_2 mixed in to increase the production of chemically reactive atoms and molecules, most of the reactive species is produced in the region where the plasma pulses are in contact with the ambient air. Thus the biochemical reaction process is probably not the major biocidal mechanism of APPJs.

APPJs are able for extremely localized treatments (down to dimensions of living cells); on the other hand, larger surface treatment may be possible by joining many such jets or by multi-electrode systems. The distance between the device and the treated sample (such as the tissue/skin) is to a certain degree variable, because it does not use the treated sample as the ground electrode, significantly simplifying treatment of the patient. Low temperature plasma jets have been used in various biomedical applications ranging from the inactivation of bacteria to the killing of cancer cells.

2.3 Arc Seeded Microwave Plasma (ASMP)

Microwaves can provide an electrodeless discharge to produce a relatively large volume plasma with no need of gas flow through the discharge. However, use of microwaves to produce atmospheric pressure air plasma in a chosen open region away from the source has to avoid the undesirable ionization along the propagation path before reaching the preferred ionization region. The undesirable ionization causes attenuation of microwave power, which usually will become too low to cause air breakdown in the designated region.

A tapered rectangular cavity is used to couple the microwaves to the exit of an arc discharge module, where the plasma is generated. An S-band tapered rectangular cavity is used to store and couple microwaves to the arc discharge module. This cavity consists of three sections. Sections on two sides have uniform cross sections "a $\times$ b" and "a $\times$ b'" with the same width "a", and different heights, "b" and "b'". One (with a cross section of a $\times$ b) is non-tapered having a length of $3\lambda_z/8$, and the other one (with a cross section of a $\times$ b') is tapered with a length of $\lambda_z/2$. The middle transition section has a length of $\lambda_z/2$ to reduce the height of the tapered section to a desirable value b' (from the height b of the non-tapered section). λ_z represents the wavelength in the axial direction.

Two aligned holes on the bottom and top walls at an electric field maximum location in the tapered section of the cavity are introduced (i.e., at $\lambda_z/4$ distance from the end-wall). A gas plenum chamber aligned to the holes is welded to the bottom wall of the cavity. It is used to feed the gas flow through, as well as to host the arc discharge module generating the seeding plasma for the subsequent microwave discharge. The arc discharge module is then screwed into this plenum chamber to the bottom hole, and the top hole allows the plasma to stream out of the cavity, as shown in Fig. 2.4a.

Because the arc discharge is contained inside the cavity, even using ambient air (rather than noble gases) as the working gas, the generated plasma has acceptable low temperature in disinfection and decontamination applications.

Microwaves generated by a magnetron (2.45 GHz, 700 W average power) radiates into this cavity at a location about a quarter wavelength ($\lambda_0/4$) (more precisely, $\lambda_z/8$) away from the shorted-end of the non-tapered section of the cavity, where $\lambda_0 = 122.5$ mm is the free space wavelength and $\lambda_z = \lambda_0\,[1-(\lambda_0/2a)^2]^{1/2} = 233$ mm is the axial wavelength for the TE_{103} mode; $a = 72$ mm is the dimension of the wider side of the cross section.

Both the arc discharge module and magnetron are run in 60 Hz with less than 50% duty cycle. The synchronization of the running phases of the two components in each cycle is essential to the operation of this setup. The optimal operation condition is when the arc discharge pulse of the module and the microwave pulse of the magnetron overlap. This can easily be attained when the two components share the same power supply. The power supply and the electric circuits used to light the discharge module and to run the magnetron are shown in Fig. 2.4b. A single power transformer with a turns ratio of 1:17 is used to step up the 60-Hz line voltage of 120 V (rms; i.e., ~170 V peak) to 2 kV (rms; i.e., ~2.83 kV peak), which is applied to both components through two series connected 1 μF capacitors,

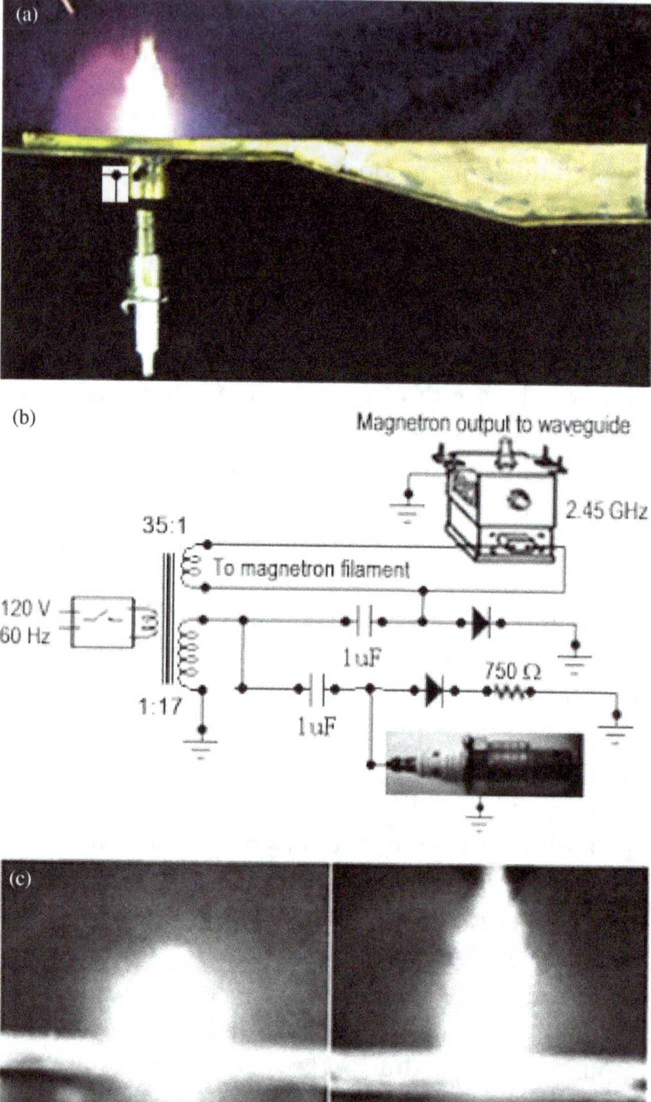

Fig. 2.4. (a) A photo showing where to insert a discharge module into a tapered microwave cavity and showing the generated microwave plasma effluent, (b) a schematic of the power supply running a module and a magnetron together, and (c) plasma plume images without airflow (left) and with airflow at a rate of 71 slm (right).

one for each component. A diode (15 kV and 750 mA rating) is used to prevent the positive voltage from being applied to the magnetron cathode. Such a half-wave rectifier elevates the negative voltage applied to the magnetron cathode to exceed the required operating voltage (~ -4.2 kV) of the magnetron.

Thus the module electrodes are also connected in parallel to a diode which in addition is connected in series with a resistor (750 Ω). This added circuit increases the voltage applied to the electrodes when the diode is reverse biased; it allows the arc discharge to occur only during the negative voltage period. The discharge pulse is shorter than the microwave pulse and the microwave field is too low to initiate discharge by itself (for the microwave power used in the present setup), it prefers to have the arc discharge to produce seeding charges right at the beginning of the microwave pulse. The series resistor delays the arc discharge to match the delay of the onset time of the magnetron due to its higher starting voltage. Thus the two discharges are synchronized as well as overlap. The microwave electric field of a TE mode is parallel to the plasma column and the evanescent microwave field extends out of the cavity hole; thus the microwaves can expand plasma generation outside of the cavity to enhance plasma volume. In other words, this new type of arc/microwave hybrid plasma generator does not need gas flow in its operation to produce sizable plasma outside the cavity. On the other hand, this discharge module adds the flexibility to introduce gas flow in the operation to increase the size of the plasma, as demonstrated by comparing the plasma plume images, in the cases of without and with airflow, presented in Fig. 2.4c; their heights are of about 15 and 30 mm, corresponding to the flow rates of 0 and 71 slm, respectively. Applied in the decontamination tests depicted in Chap. 5, the device was run at the airflow rate of 24.6 slm.

Microwave enhances significantly the arc plasma density and volume; it also increases the plasma thermal temperature. The temperature of this microwave plasma keeps it from being used to treat living tissues, but it is still low enough without causing thermal damage to instruments. It has been demonstrated that this plasma kills *B. cereus* spores, a simulant of Anthrax, rapidly and thoroughly.

Emission spectroscopy of the microwave plasma plume indicates that the device generates an abundance of atomic oxygen in the plasma effluent, which stimulates oxidation-reduction reactions to cause protein degradation as the biocidal mechanism. The applicability of this plasma for biological decontamination is discussed in detail in Chap. 5. It is also adaptable for most of the sterilization applications.

2.4 Air Plasma Spray (APS)

A discharge module, which consists of a pair of concentric electrodes, is used as plasma generator. At atmospheric pressure, the air discharge between the circular electrodes is an arc mode, which generally evolves into a constricted arc and develops hot spots on the electrode surfaces. The airflow and the magnetic field are introduced in the device in order to prevent the discharge from forming arc constriction and hot spots.

The airflow pushes the arc between the electrodes into an elongated loop, which can extend out of the gap between electrodes more than 30 mm. It increases the discharge path length and reduces the transit time loss of the discharge significantly, which have the effect of reducing the arc temperature, and the sparkling and heating of the electrodes, and of increasing the plasma volume and the generation region of the reactive species, which is particularly important to those with short lifetimes, such as atomic oxygen.

A concentric ring magnet is installed behind the outer electrode. In the electrode gap region, this ring magnet produces a magnetic field $\mathbf{B} = B_z \hat{z} + B_r \hat{r}$ having axial and radial components, which exerts a force $\mathbf{F} = \mathbf{J} \times \mathbf{B}$ on the current density $\mathbf{J} = J_z \hat{z} + J_r \hat{r}$ of the discharge, in the azimuthal direction. This force rotates the discharge loops to prevent the formation of hot spots on the electrodes. Thus the discharges are maintained in the elongated arc loop which rotates into a diffused cone over time. The magnetic field stabilizes the discharge, slows down the build-up of the arc temperature, and reduces arc erosion on the electrodes.

These are essential arrangements to keep the discharges stable, and to expand the volume such that a sizable non-thermal plasma

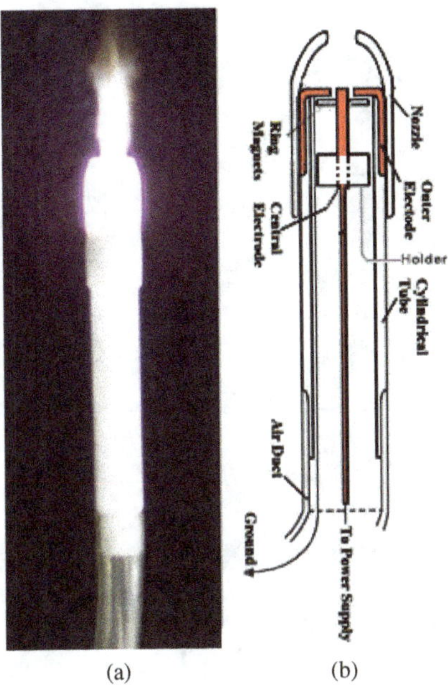

Fig. 2.5. (a) Photo of a handheld air plasma spray in running, and (b) a schematic of the spray.

plume protruding through the gap of the electrodes is generated. This is demonstrated in Fig. 2.5a, in which plasma is generated by a handheld air plasma spray using a blower to provide air flow and placing a ring magnet right under the outer electrode. A schematic of the air plasma spray is also shown in Fig. 2.5b.

A cap (nozzle) is introduced to direct the flow of the plasma effluent as well as to cover the electrodes for safety, so that the high voltage (HV) central electrode is not exposed. The size of the discharge module can vary with the specific application.

The gap between the electrodes is in the range of 0.8 to 1 mm. The ring-shaped magnet produces a magnetic field of ~0.18 T (Tesla) in the electrode gap region. An airflow rate of ~21 slm is applied. Air plasma spray is run in periodic discharge mode; the power supply is shown in Fig. 2.6a, which is similar to that used to run the arc discharge module of the ASMP, except that the 750 Ω resistor in series with the diode is removed. The produced plasma

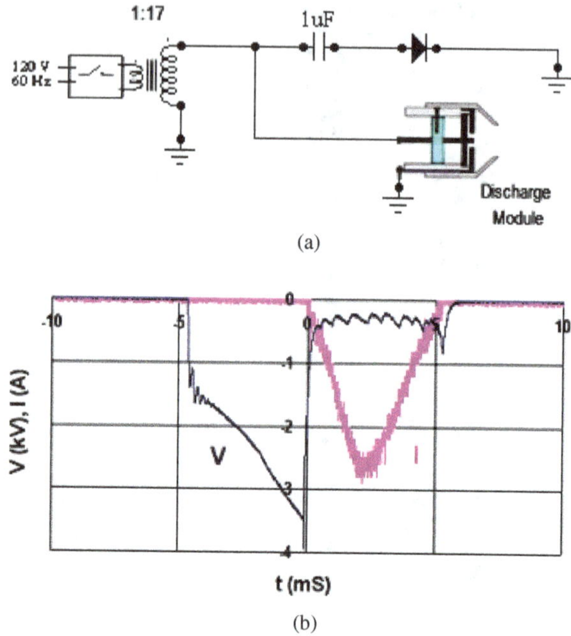

Fig. 2.6. (a) A circuit diagram of the power supply, and (b) voltage and current functions of the discharge in one cycle.

effluent protrudes from the exit of the cap more than 30 mm (i.e., more than 40 mm away from the discharge gap).

The turns ratio of 1:17 of the power transformer enables it to step up the 60-Hz line voltage of 120 V (rms) to 2 kV (rms), which is rectified by a diode and magnified by a series connected 1µF capacitor to exceed 4 kV level. Due to the concentric cylindrical configuration of the electrodes, the electric field intensity near the central electrode is much stronger than that near the outer electrode; air breakdown is usually triggered in the region near the central electrode. Thus, it is preferable to apply negative voltage to the central electrode so that the electrons generated in the air breakdown will not be quickly collected by the central electrode.

Air breakdown occurs when the voltage applied to the central electrode of the discharge module reaches about −3.5 kV. In each cycle of the 60 Hz AC input, discharge occurs only in the negative half cycle due to this rectifier diode used in the circuit.

The time varying voltage V(t) and current I(t) of the discharge were measured using a digital oscilloscope (Tektronix TDS3012 DPO 100 MHz and 1.25 GS/s), where V is the voltage of the central electrode of the module (the outer electrode is grounded). These two time functions, V and I, in one cycle, are presented together in Fig. 2.6b, to show their phase relationship. As shown, after the breakdown occurs, the voltage drops rapidly and the discharge is basically sustained by a constant voltage about −320 V. The duty cycle of the discharge is about 25%. The peak power in each discharge is about 800 W and the average power is about 100 W.

The thermal temperature (<340 K) of the plasma effluent is much lower than the excitation temperature of electrons channeled by the rotating arc loop, which is estimated via emission spectroscopy to be higher than 7700 K. It explains why APS carries an abundance of atomic oxygen to be shown in Chap. 3.

This plasma spray is non-equilibrium due to the following factors associated with the design and the operation, 1) the discharge is run in a periodic mode with a low duty cycle, 2) the airflow pushes the discharge current to increase the size of the arc loop considerably and cool the arc and electrodes, 3) the magnetic field rotates the arc discharge, preventing the formation of arc constriction and hot spots on the electrode surfaces, 4) the discharge is run in relatively high voltage/relatively low current mode (e.g., relative to the dc arc discharges powered by a current source). Because the discharge extends 30 to 35 mm beyond the discharge gap, a considerable number of electrons are generated locally in the plasma plume, rather than being conveyed from the gap region of the electrodes. In-vitro and in-vivo (using pigs as the animal models) tests, presented in Chaps. 6 and 7, have verified that APS can rapidly clot blood to stop wound bleeding.

The power transformer in the electric circuit, shown in Fig. 2.6a, is heavy and is not efficient when running a dynamic load. The power supply also overruns the discharges, which elevate the plasma temperature. However, as shown in Fig. 2.6b, the maintaining voltage of the discharge after air breakdown is less than 350 V. It suggests that the discharge module can be run with a power supply designed to consist of an ignition coil, which generates high voltage dc pulses (e.g., ~20 kV) to trigger air breakdown, and a low voltage pulsed dc power supply (e.g., ~350 V or less), which sustains low voltage pulsed discharge following each air breakdown. Based on this understanding, a new power supply has been designed to run the air plasma spray. An essential requirement of the circuit design is that the low voltage discharge pulse has to synchronize with the high voltage trigger pulse, and is preferable to start right after the trigger pulse. The block diagram of this new power supply will be presented in Chap. 9. Bleeding control of APS run with this new power supply were examined by in-vivo trials using pigs as the animal models; the results are also presented in Chap. 9.

The experimental trials successfully demonstrate that, for an amputated wound, the hemorrhage was controlled in a few minutes and the tourniquet could be removed right after the treatment. In other words, APS treatment is able to reduce tourniquet-use time and to downgrade tourniquet-necessary wound. It meets the golden period of saving the remaining part below the tourniquet. The consequence of plasma treatment on wound healing was also demonstrated. In post-operative observation of the healing of the cross cut wound, a single plasma treatment to stop bleeding also reduces the wound healing time to about half. These positive results encourage the development of APS for advanced first aid, which is discussed in detail in Chap. 9.

Problems

P2.1. A parallel-plate capacitor of width w, length L, and separation d is partially filled with two dielectric slabs of dielectric constant ε_r with different thickness d_1 and d_2, respectively. As shown in Fig. P2.1, these two dielectric slabs fill the regions $0 < z < d_1$ and $d-d_2 < z < d$ to cover the two conducting plates of the capacitor at $x = 0$ and $x = d$. A battery of V_0 is connected between the two plates.

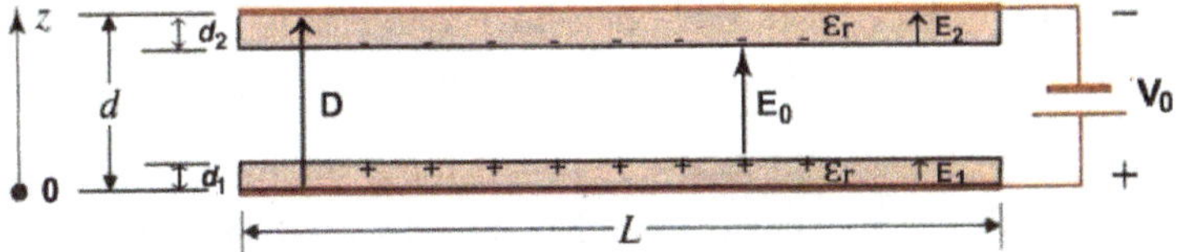

Fig. P2.1. A partially filled capacitor.

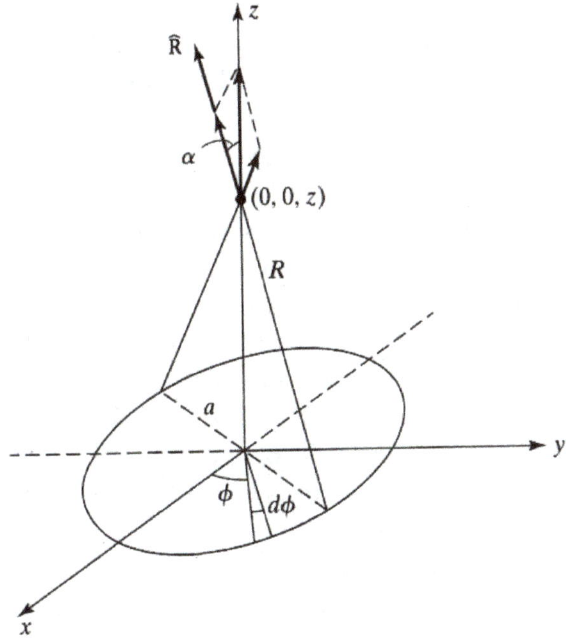

Fig. P2.2. Coordinates and orientation of a charged circular ring.

(1) Find **D** and **E** in each region.
(2) Find ρ_s at x = 0, d_1, $d-d_2$, and d.
(3) Total charge on each plate of the capacitor and on each interface surfaces

P2.2. Charge Q is uniformly distributed along a circular ring of radius a lying in the xy-plane and having its center at the origin, as shown in Fig. P2.2. Find the electric field intensity at a point on the z-axis.

P2.3.   We now place a similar circular charge ring of $-Q$ above the circular charge ring of Q, in Problem P2.2, at $z = d$.

    (1)   Find the electric field intensity at a point on the z-axis.

    (2)   Find the location of the maximum field intensity, and the value of the maximum field intensity.

    (3)   Find the d value to obtain maximum electric field intensity at $z = d/2$ for a fixed V_0.

P2.4.   Consider a SDBD APPJ as shown in Fig. 2.2b; each of the ring electrodes has a width "w" and the quartz tube has a wall thickness of "δ", inner radius of "a", and dielectric constant ε_r. Find the induced charge on the inner surface of the quartz tube.

P2.5.   The APS shown in Fig. 2.5 is generated by arc discharges, which erode the electrodes by the bombardment of energized charge particles. In the electrode setup of the APS device, the central electrode has much less surface area and thus is more vulnerable to the damage of the bombardment. Based on the analysis of Problem P1.1, explain why it is preferable to connect the central electrode of the generator to the negative voltage terminal of the power supply to reduce the erosion of the electrode. (In ion-neutral collision, $\delta = \frac{1}{2}$; and $v_{in}/v_{en} \approx (m_e/m_i)^{1/2}$, where v_{in}/v_{en} are the ion/electron-neutral elastic collision frequencies and m_e/m_i are the electron/ion masses.)

P2.6.   Set up an Air Plasma Generator:

    (1)   Set up a power supply with the circuit diagram shown in Fig. 2.6a, in which the transformer, diode, and capacitor in the circuit are the same as those used in a similar circuit shown in Fig. P2.3a for a conventional microwave oven (i.e., non-inverter type). Thus, an easy way is to use the power supply of a microwave oven. Purchase a cheapest microwave oven and open it in the back to extract the power supply.

         Compare Fig. 2.6a and Fig. P2.3a, the output terminals of the transformer, which connect to magnetron filament, will not be used any more (leave those alone);

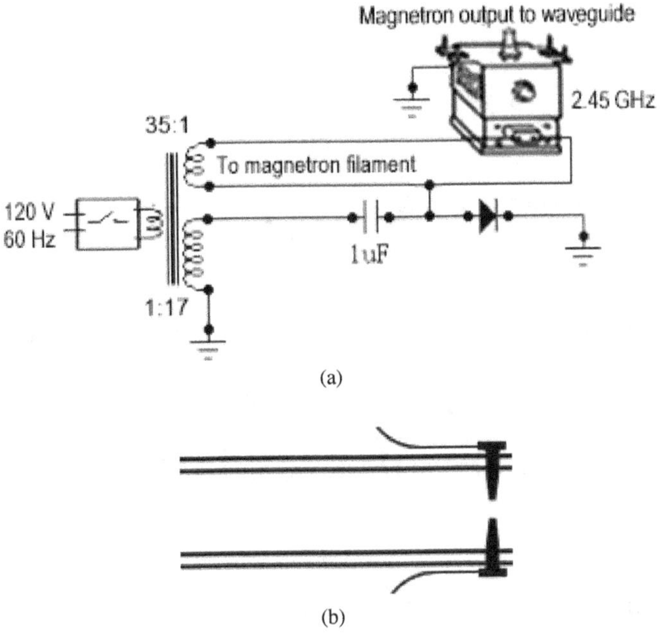

Fig. P2.3. (a) Circuit diagram of the power supply of a microwave oven, and (b) a schematic of a plasma generator.

and a wire, which is connected to the capacitor and disconnected from the magnetron, will be the high voltage output terminal of the power supply.

The body of the transformer is grounded, and an electric wire can be connected to one of the post screws to setup a ground terminal.

(2) Set up the generator.

Get a piece of PVC pipe of small diameter (1 inch or less) and two copper screws (not brass). Make two aligned holes in PVC tube for two screws to insert through as shown in Fig. P2.3b, where the gap between the screws is adjusted to be around 2 mm and the tip of each screw is sharpened. The gap is adjusted to allow the breakdown to occur.

(3) Connect the other end of the PVC pipe to an air pump (an air mattress blower or a small axial turbo fan).

Chapter 3

Free Radicals and Reactive Oxygen Species in Cold Atmospheric Plasma

3.1 Biocidal Effects of Free Radicals and Reactive Oxygen Species

Reactive oxygen species produced by macrophages and neutrophils of the mammalian immune system cells, play a fundamental role in the antibacterial and antiviral defense; platelets involved in wound hemostasis and repair also release ROS to recruit additional platelets to sites of injury; thus exogenous reactive oxygen species, such as superoxide radicals, hydrogen peroxide, singlet oxygen, hydroxyl radicals, and atomic oxygen generated by plasmas, are expected to provide similar functions in bacterial inactivation and blood clotting.

Radicals and reactive oxygen species (ROS) can wreak havoc on a broad range of macromolecules. A prominent feature of radicals and ROS is that they have extremely high chemical reactivity, which inflicts damage on cells. Membrane breaching by ROS has been substantiated; this could be caused by either physical punctures or oxidation of membrane proteins, which trigger an irreversible cell damage; on the other hand, DNA damage has been perceived as less important.

Those species can be generated through electron impact excitation and dissociation. Thus non-equilibrium gas plasmas can carry an abundance of radicals, responsible for the inactivation of microorganisms in the applications of plasma disinfection and decontamination. However, it is noticed that the reactive plasma species under atmospheric conditions has a relatively short lifetime; as a result, most of the plasma active species, with the exception of the metastable ones (long-life electronically excited atoms), drop their fluxes dramatically in a short distance away from the source. There are many types of radicals, which depend on the working gases and plasma generation methods. Noble gases such as Helium (He) and Argon (Ar) and their mixtures with oxygen (O_2), inert gas such as nitrogen (N_2) and its mixture with oxygen (O_2), and ambient air have been used as working gases.

Free radicals carried in plasmas generated in noble and inert gases are not oxidizing agents and do not instigate oxidation–reduction (redox) reactions. In many tests with atmospheric pressure argon plasma jets, inactivation of spores was found to be mainly attributed to UV radiation. Although low density ROS may be generated when the plasma jet encounters the ambient air outside the device, the mechanisms of actions of such plasmas are speculated to be mainly using electrons, ions, atoms and molecules in the excited states, in addition to the UV radiation, to attack microbial structures, specifically the inner membrane and DNA in the core of bacterial endospores. However, the inner membrane is covered by a coat, it is accessible only for the most energetic ones in the plasma. As a result, the penetration rate is low; thus the kill rate through this mechanism is much lower than that through chemical means, such as oxidation-reduction reactions which oxidize the outer structures of cells by the ROS. This is indeed the case of experimental observation.

Atmospheric pressure helium plasma jets generated with and without oxygen were used in disinfection tests; it was observed that the D-value (the time needed to kill 90% of the microorganisms) was higher (i.e., killing rate was lower) when oxygen was absent. Adding oxygen to the helium gas discharge made the device more effective at killing bacteria (i.e., decrease of the D-value). For example, a plasma pencil device, running with three different working gases, pure helium, a mix of 97% helium/ 3% of oxygen, and air, was used to deactivate Bacillus spores. Pure helium plasma gave a

D-value of over 20 minutes; without oxidizing agents in the plasma to cause surface damage, the exterior structure of sterilized spore remained almost unchanged. Adding 3% oxygen to the helium decreases the D-value to 10 minutes. When helium was replaced by air as the working gas, the D-value decreased abruptly to 20 seconds. It is concluded that moistened oxygen and air as working gases gave the best plasma bactericidal effects (i.e., with the lowest D-value).

This is because reactive oxygen species (ROS), which are those of most concern in biological systems, could be generated in oxygen-based gas plasmas, such as air plasma, through electron impact excitation, attachment, and dissociation. The likely radicals include excited oxygen such as singlet oxygen ($^1O_2$), superoxide ($*O_2^-$), atomic oxygen (OI), and ozone (O_3), etc. Ozone has good bactericidal capacity, but its concentration in the CAP has to be restricted in the clinic treatment due to its toxic property. It is shown later (see Fig. 5.5) that the D-value of *B. cereus* spores treated by ASMP, which carries an abundance of OI, is less than 5 seconds. Indeed, the atomic oxygen plays the main role in the applications. OI atoms are capable of abstracting hydrogen from the protein polymer backbone; those reactions create radical sites to initiate subsequent processes that lead to the cleavage of the polymer chain and subsequent reaction with oxygen atoms or molecules, causing irreversible protein degradation.

The cell membrane is composed mainly of lipids, proteins, and polysaccharides. Lipids are macromolecules that are susceptible to ROS attack *via* the H-abstraction mechanism. Protein molecules, which are basically linear chains of amino acids, are vulnerable to oxidation by ROS. Hence, ROS, in particular OI atoms, can compromise the barrier function of the membrane. Oxidation-reduction (redox) reactions can start at cell surface; the oxidized proteins are damaged irreversibly, leading to the loss of cell viability. Consequently, ROS give much faster kill rate through oxidation processes, unnecessary to impose DNA damage.

ROS can also trigger natural, rather than thermally induced, blood coagulation cascades. CAAP treatment activates the coagulation cascade to enhance the aggregation of platelets and the conversion of the coagulation protein fibrinogen into fibrin. Fibrin strands then strengthen the platelet plug to form a hemostatic plug. Plasma agents selectively trigger the polymerization of coagulation proteins, without altering the albumin structure. Moreover, CAAP can regulate angiogenesis via stimulating or inhibiting cell proliferation.

There are four likely processes to produce atomic oxygen; one is to dissociate an oxygen molecule into two oxygen atoms via the reaction $e^- + O_2 \rightarrow 2O + e^-$, which has a reaction rate coefficient $k_1 = 4.2 \times 10^{-9} \exp(-5.6/T_e)$, where T_e is in eV. This reaction rate decreases rapidly with $T_e < 5.6$ eV; thus it needs about 5 eV electrons to effectively dissociate O_2 into atomic oxygen. The second one is through recombination of charged particles $e^- + O^+ \rightarrow O$, which may not need the presence of energetic electrons. The third one is the dissociative attachment of electrons to molecular oxygen $e^- + O_2 \rightarrow O + O^-$; conservation of energy requires that the electron energy exceeds a threshold level, which normally is 3.6 eV. Though this threshold level decreases as the internal energy of molecular oxygen increases, laboratory experimental results show that this level can go down to 1 eV and less. The fourth one is the dissociative recombination reaction $O_2^+ + e^- \rightarrow O + O$. In the air discharge, where oxygen has lower ionization energy than that of the nitrogen, electrons and O_2^+ are produced. As the air plasma flows out of the discharge zone, atomic oxygen (OI) is generated in the plasma effluent via this dissociative recombination process. Therefore, the fourth process of dissociative recombination is likely the dominant process of atomic oxygen generation in low temperature air plasma.

3.2 Plasma-liquid Interaction to Generate Additional Free Radicals and Reactive Oxygen Species

Additional ROS are generated when these radicals react with water embodied in the treated subjects. For example, water (H_2O) is converted to hydroxyl radical (*OH) and hydrogen peroxide (H_2O_2), via reacting with atomic oxygen of different states,

$$O(^3P) + H_2O \rightarrow 2 \, {}^*OH$$

and

$$O(^1D) + H_2O \rightarrow H_2O_2$$

Hydrogen peroxide is a reactive oxygen species, but itself is not a free radical as it does not contain any unpaired electrons. However, it is a precursor to certain radical species such as atomic oxygen, peroxyl radical, hydroxyl radical, and superoxide. Since it is a superb oxidizing agent it can also react with certain other molecules and convert them into free radicals.

Most strains of harmful bacteria (and cancer cells) are anaerobic and cannot survive in the presence of oxygen or H_2O_2. Our bodies create reactive oxygen species (ROS) and use free radicals to destroy harmful bacteria, viruses, and fungi. In fact, the cells responsible for fighting infection and foreign invaders in the body (white blood cells) make hydrogen peroxide (H_2O_2), which releases a free radical atomic oxygen (OI) to oxidize any offending culprits. H_2O_2, produced within individual body cells, is essential for life.

Hydrogen peroxide (H_2O_2) stimulates enzyme systems throughout the body. This triggers an increase in the metabolic rate, causes small arteries to dilate and increase blood flow, enhances the body's distribution and consumption of oxygen and raises body temperature.

Hydrogen peroxide is used for important cell signaling functions. The healing of skin wounds restores not only the integrity of the skin, but also its sensory function. Following injury, skin cells proliferate and migrate to seal the wound, and peripheral sensory axons innervating the skin must also regenerate to restore sensory function. Hydrogen peroxide released by wounded skin cells promotes the regeneration of sensory fibers, thus helping to ensure that touch sensation is restored to healing skin.

However, if its concentration gets too high, highly reactive hydroxyl radical (*OH) can be produced, e.g., via UV-light dissociation of H_2O_2. On the other hand, catalase catalyzes the decomposition of hydrogen peroxide to water and molecular oxygen. This serves to modulate levels of peroxide in the body.

Unlike superoxide, which can be detoxified by superoxide dismutase, the hydroxyl radical cannot be eliminated by an enzymatic reaction, as this would require its diffusion to the enzyme's active site. However, the hydroxyl radical has a very short in vivo half-life of ~10^{-9}s. As diffusion is slower than the half-life of the molecule, it will react with any oxidizable compound in its vicinity and also immediately remove electrons from any molecule in its path, turning that molecule into a free radical and thus propagating a chain reaction. The hydroxyl radical can damage virtually all types of macromolecules, carbohydrates, nucleic acids (mutations), lipids (lipid peroxidation) and amino acids (e.g. conversion of Phe to m-Tyrosine and o-Tyrosine). When the rate of the cell wall damage is higher than the rate of its own reparation, it leads to the death of the cell. In the emission spectroscopy, the *OH band at 306.4 nm is usually buried in the metallic lines.

When Ar is used as the working gas in plasma diffusing into an aqueous liquid medium, *OH radicals can be generated by charge exchange of water vapor with argon ions Ar^+

$$Ar^+ + H_2O \rightarrow H_2O^+ + Ar$$

which leads to

$$H_2O^+ + H_2O \rightarrow OH + H_3O^+ \text{ and } H_3O^+ + e^- \rightarrow OH(A) + H_2$$

and by dissociative excitation of water vapor with metastable argon Ar^*, excited by electrons: $Ar + e^- \rightarrow Ar^* + e^-$, as follows

$$Ar^* + H_2O \rightarrow Ar + OH(A) + H^*$$

Similarly, when Helium plasma diffuses into an aqueous liquid medium, *OH radicals are generated by dissociative excitation of water vapor with metastable helium He^*, excited by electrons: $He + e^- \rightarrow He^* + e^-$, as follows

$$He^* + H_2O \rightarrow He + OH(A) + H^*$$

On the other hand, hydrogen peroxide is actually more damaging to DNA than the hydroxyl radical. This is because the lower reactivity of hydrogen peroxide provides enough time for the molecule to travel into the nucleus of the cell, subsequently reacting with macromolecules such as DNA. Hydroxyl radicals may combine to hydrogen peroxide via the reaction

$$*OH + *OH \rightarrow H_2O_2$$

In the case that plasma is generated in liquid, such as by electric discharge in water (for water purification), OH radicals are generated directly via dissociative electron excitation of water

$$e^- + H_2O \rightarrow OH(A) + H + e^-$$

and via dissociative recombination of water ions

$$e^- + H_2O \rightarrow H_2O^+ + 2e^-$$

$$H_2O^+ + H_2O \rightarrow OH + H_3O^+$$

$$H_2O^+ (H_3O^+) + e^- \rightarrow OH(A) + H_{(2)}$$

Electron impact with a minimum energy 15.7 eV on water also produces excited oxygen atom $OI(3p^5P\text{-}3s^5S^0)$ and $OI(3p^3P\text{-}3s^3S^0)$, which can directly react with water molecules to generate intermediate species such as hydroxyl or hydrogen peroxide to cause further inactivation of bacteria.

The atomic oxygen and hydroxyl radical play effective roles in microorganism inactivation. Hydrogen peroxide adds synergistic effects with the long-term inactivation property (high initial concentration of H_2O_2) and high antibacterial efficiency (low initial concentration of H_2O_2). These efficacies are the basis of plasma biomedicine, food decontamination, and water purification applications.

3.3 Emission Lines of Likely Reactive Species in CAAPs

The speed "c" of light (electromagnetic radiation) and its frequency "f" and wavelength "λ" are related by

$$c = \lambda f \tag{3.1}$$

where $c = 3 \times 10^8$ m/s; the frequency f is in cycles per second (i.e., Hertz or s^{-1}); the wavelength λ is in meters (m) or in nanometers (nm); 1 nm $= 10^{-9}$ m.

Each photon in the radiation carries energy

$$E = hf = \frac{hc}{\lambda} \tag{3.2}$$

where $h = 6.624 \times 10^{-34}$ joule seconds (J-s) is the Planck's constant; the photon energy E is in joules (J) or in electron volt (eV), here, $1\ eV = 1.6 \times 10^{-19}$ J.

The wavelength of a photon emitted by an atom or a molecule during its transition from an excited state to a lower energy state is inversely proportional to its energy (i.e., $\lambda = hc/E$), which is determined by the difference of the energy levels (i.e., energy gap) of the two states involved in the transition. Each element in a material emits a characteristic set of discrete wavelengths according to its electronic structure, a spectrometer is used to separate the components of the emission, which have different wavelengths. Thus, the elemental composition of a material can be identified via emission spectroscopy. The emission spectrum of the material over a wavelength range is recorded, in which the wavelength of the spectral line gives the identity of the element while the intensity of the spectral line is proportional to the concentration of the element.

In CAAPs, the likely reactive species are ROS and RNS which include atomic oxygen (OI), singly ionized oxygen molecule (O_2^+), singlet oxygen ($^1O_2$), ozone (O_3), excited nitrogen (N_2^*), singly ionized nitrogen (N_2^+), and nitric oxide (NO). In addition, *OH radical could also be produced. Some of these emit discrete lines having easily identifiable spectral spikes, but others have emission lines distributed in a band with a band head. A spectrometer can scan a wide band to reveal the spectral peaks of emission from the sample, but it may miss weak lines, in particular, for those distributed in a band. This can be remedied by scanning only a narrow band around the characteristic lines of each element. Following are the characteristic lines of each likely reactive species:

(1) Most intensive emission lines of atomic oxygen (OI) include these at

$$777.194, 777.417, \text{ and } 777.539 \text{ nm}; 770.68 \text{ nm};$$
$$844.64 \text{ nm}; 615.819 \text{ nm}; 394.729 \text{ nm}; 436.83 \text{ nm};$$
$$\text{and } 202.64 \text{ nm}.$$

The quantum efficiency is much higher at 777 nm than at other wavelengths. Therefore, the 777 triplet (777.194; 777.417; 777.539) in the emission spectroscopy is used as the atomic oxygen actinometry, estimating OI concentration.

(2) Most prominent band heads of singly ionized oxygen molecule (O_2^+) include these at

$$298.6 \text{ nm and } 304.5 \text{ nm}.$$

(3) Singlet oxygen ($^1O_2$) studies are usually performed by steady-state and time-resolved phosphorescence measurements with emission detection around 1270 nm. Such measurements are usually challenging, because the singlet oxygen emission is very weak compared to, e.g., the fluorescence signal of the photosensitizer.

(4) Band heads of excited nitrogen (N_2^*) include these at

$$315.93 \text{ nm}; 337.13 \text{ nm}; 357.69\,(v' = 0, v'' = 1) \text{ nm};$$
$$375.55 \text{ nm}; 399.84 \text{ nm}; 427 \text{ nm}; 750.3 \text{ nm};$$
$$\text{and } 782.85\,(v' = 2, v'' = 2) \text{ nm}.$$

(5) Emission lines and band heads of singly ionized nitrogen molecule (N_2^+) include these at

$$388.43 \text{ nm}; 391.44 \ (v' = 0, v'' = 0) \text{ nm};$$
$$427.81 \ (v' = 0, v'' = 1) \text{ nm};$$
$$443.3 \text{ and } 444.7 \text{ nm}; \text{ and } 463 \text{ nm}.$$

(6) Band heads of Nitric oxide (NO) include these at

$$218.18 \text{ nm}; 222.24 \text{ and } 223.61 \text{ nm}; 237.02 \text{ nm};$$
$$\text{and } 247.87 \text{ nm}.$$

(7) The emission spectrum of OH can be observed at

$$306.36 \text{ nm}.$$

(8) The most important ozone absorption spectral region is the Hartley band, extending from slightly above 300 nm down to

slightly above 200 nm. It is this band that is responsible for absorbing UV C in the stratosphere. Because there is a very intense maximum absorption at 255 nm with absorption cross section about 10^{-17} cm^2 per molecule, one can use a mercury-vapor lamp to transmit emission at 253.7 nm through a CAP to explore O_3. The intensity drop of the 253.7 nm line is proportional to the concentration of ozone in the CAP. The UV absorption method for Ozone density (concentration) N_{O3} measurement is performed as follows:

The 253.7 nm UV radiation generated by a mercury-vapor lamp is illuminated into the plasma effluent through a slit with a hole (e.g., a diameter of 1mm) and is detected by a spectrometer through a second slit with the same size hole.

N_{O3} is then calculated using Lambert–Beer's law:

$$N_{O_3} = -\left(\frac{1}{\sigma L}\right)\ell n\left(\frac{I}{I_0}\right) \qquad (3.3)$$

where I and I_0 are the detected light intensities in the presence/absence of the plasma effluent, respectively; L is the effective optical length and σ is the cross section of O_3 against a 253.7 nm photon: 1.15×10^{-17} cm^2.

3.4 Emission Spectroscopy of Atomic Oxygen

Atomic oxygen flux of the plasma can be examined via the emission spectroscopy of the plasma plume. This is described in the following. An air plasma spray (APS) shown in Fig. 3.1a, which was run by the power supply shown in Fig. 2.6a, is used to illustrate the study of the emission spectroscopy. The visible plasma plume extends out axially to about 40 mm from the nozzle of the spray. The emission spectrum of the plasma plume outside the nozzle from 300 to 900 nm was scanned by a spectrometer. In this spectral range, the UV radiation from 300 to 400 nm was not detected and the intensive lines contributed by oxygen radicals appear only around 777.4 nm, which were emissions from the 5P state of atomic oxygen (OI) in the plasma effluent of the spray. A 2-D measurement of the spatial

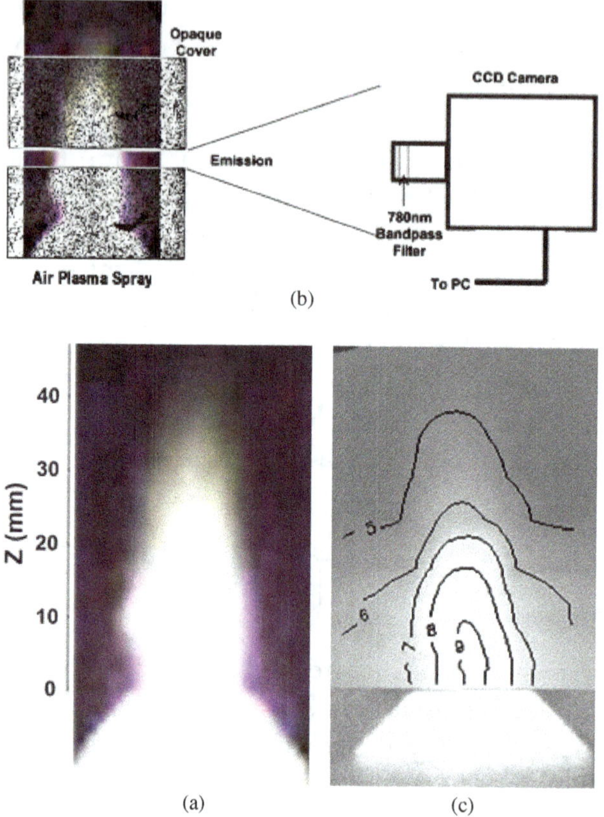

Fig. 3.1. (a) A video graph of APS, (b) a schematic of the measurement setup, and (c) the spatial distribution of the emission intensity at 777.4 nm of the APS; the numerical value at each contour is the $\log_{10}$ of the intensity in Rayleigh.

distribution of 777.4 nm radiation intensity is described in the following.

Shown in Fig. 3.1b is a schematic of the measurement setup. A 50 mm diameter 10 nm pass band interference filter with a center wavelength of 780 nm was used to focus on measuring the spatial distribution of the 777.4 nm lines. The choice of lens relative to detector size ensures that the target emission line passes through the filter over the entire field of view; thus the image data from the entire plasma plume could be taken simultaneously with an Apogee

Alta E47 thermoelectrically cooled CCD camera with a Nikon 50 mm f/1.2 lens. Another advantageous feature of the camera measurement is that, for a diffuse pixel-filling source such as the plasma plume, the distance between the camera and the source does not change the measured intensity per pixel.

The camera was calibrated to determine the bias counts, thermal noise, and sensitivity by acquiring images of an integrating sphere calibration source traceable to the National Institutes of Standards and Technology at the exposure times and aperture settings used in the experiment. Intensity in the unit of Rayleigh (apparent emission rate of 10^{10} photons/m^{2}s integrated along a line of sight) was determined from a calibration table provided with the source and the filter transmission documented by the manufacturer. This calibration provided a linear relationship between pixel counts and the intensity of 777.4 nm emissions in the plasma plume for each aperture setting and exposure time. The volume emission rate in photons m^{-3}s^{-1} can be determined by dividing the apparent intensity in Rayleigh by the length of the path through the emitting region, in this case the path length is approximately 10 mm.

Due to the short lifetime (~0.2 to 0.3 ms) of atomic oxygen, the axial distribution of the emission intensity was expected to vary strongly and to have a sharp upward extent. To capture the full axial extent of this distribution beyond the dynamic range of a single 16-bit digital image, multiple images of the APS were recorded separately with a moveable stage approximately 0.15 m square draped in black velvet cloth blocking the brighter portions of the APS to allow operation of the camera at higher sensitivity and longer exposure times while avoiding saturation in the brighter parts of the plasma plume. Due to the size of the barrier and the diffuse covering, the measurements were made in the geometrical optics regime with diffraction negligible relative to the resolution of the camera.

The imaging process started by blocking most of the flame region to detect the upward boundary of the emission using the camera at high sensitivity with the aperture opened to f/1.2 and the exposure time set at 1 s. The cover was then moved down by 6.4 mm (a quarter inch) sequentially in each image, with the aperture and exposure time adjusted accordingly at each stage. With the barrier completely lowered, the bright emissions from the direct flames required the camera to be operated at its lowest sensitivity, with the aperture stopped to f/16 and exposure time reduced to 0.1 s. Even at this shortest exposure time of 0.1 s, the image was an average over 6 discharges. In other words, the instantaneous fluctuation of the emission due to the 60 Hz ac driving the discharge has been evened out in all images. The resulting images were then calibrated and combined into a composite image, shown in Fig. 3.1c, giving contours of the intensity distribution, ranging from 10^5 to 10^9 Rayleighs over the axial extent of the spray. As shown, at about 40 mm from the nozzle of the spray, the intensity of 777.4 nm emissions is about 10^5 Rayleighs; i.e., the total photon emission from a slice of the spray at 40 mm away from the nozzle is about $10^{15}\mathrm{m}^{-2}\mathrm{s}^{-1}$.

Assuming that the emission is isotropic and taking into account that the plasma spray has a cylindrical shape, the corresponding flux and density of 5P state OI are estimated to be about $6 \times 10^{13}\mathrm{m}^{-2}\mathrm{s}^{-1}$ (i.e., $\sim 10^{15}/[(4\pi) \times (4/\pi)]$) and $3 \times 10^{12}\mathrm{m}^{-3}$ (i.e., $\sim 6 \times 10^{13}/20$, where the flow speed of 20 m/s is assumed), respectively.

The intensity distribution I of the 777.4-nm emissions, along the central axis (z) of the spray, is presented in Fig. 3.2, where the emission intensity I is extracted from Fig. 3.1c and $z = 0$ is set at the nozzle exit of the spray. As shown, the range of the plasma effluent emitting 777.4 nm lines extends out axially to about 40 mm from the nozzle of the spray, where the intensity of 777.4 nm emissions is about 10^5 Rayleighs.

The strong variation of the axial distribution of the emission intensity I and the sharp drop of I at near $z = 40$ mm are expected, due to the short lifetime ($\sim$0.2–0.3ms) of atomic oxygen and the low flow speed of $\sim$20m/s.

The 5P state of the transition in OI has rather high energy relative to the ground state, about 10.74 eV. Therefore, the strong line intensity outside the core of the APS (i.e., outside the white area of the contour plot) indicates that plasma is in a non-equilibrium state with a strong presence of high-energy electrons and an abundant concentration of 5P state atomic oxygen in the plasma effluent, as

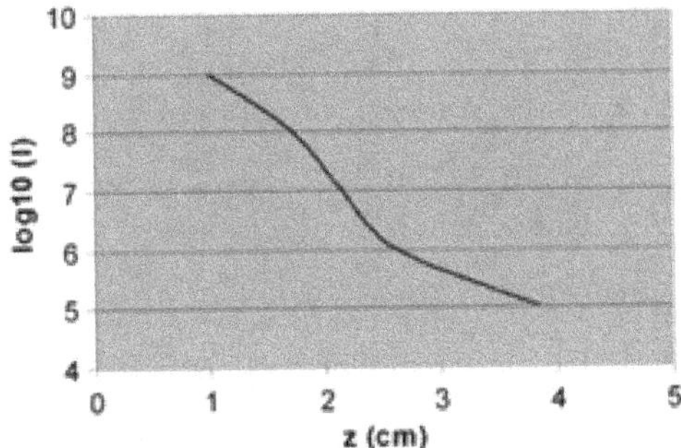

Fig. 3.2.   Spatial distribution of the emission intensity I, in Rayleigh at 777.4 nm, along the central axis (z) of the spray.

calculation assuming low optical thickness represents a minimum bound on the average density. Since OI in other states (such as ^{3}P, ^{1}D, and in particular, the ground state) are also expected to be produced, the total OI concentration in the plasma effluent should be much higher than that of 5P state alone.

3.5 Emission Spectroscopy of Electron Excitation Temperature

Shown in Fig. 3.3 is the scheme of the imaging spectroscopy used to determine axial distribution of the electron excitation temperature $T_{exc}(z)$ in APS. The image of the APS was projected to the entrance plane of the imaging spectrometer, with magnification $s = z_i/z_t = 0.08$. The entrance slit was cut through the central axis of the APS image. The CCD array detector captured the spectrally dispersed image of the irradiated slit image, which was then analyzed row by row to obtain individual spectral intensities $I(z, \lambda)$.

Given the magnification factors and the size of single pixel of 25 μm, $I(z, \lambda)$ corresponded to an average intensity radiated from a plasma cylinder 0.3 mm high and 0.1 mm wide cutting through the central axis of the APS. Each spectral frame was calibrated using a black body radiation source to eliminate the error induced by the grating position.

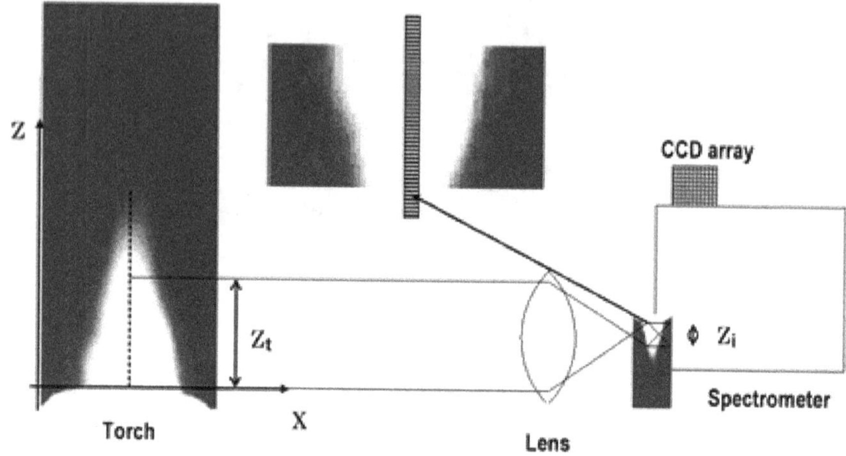

Fig. 3.3. Scheme of the imaging spectroscopy.

The recorded spectrum is abundant in Cu I lines. To determine the electron excitation temperature, we have chosen some of the lines that correspond to transition to the ground state because they are best documented. Six lines, at

216.509 nm, 217.894 nm, 222.57 nm, 249.215 nm, 324.754 nm, and 327.396 nm,

are included in the Boltzmann plot according to the relation

$$\frac{I(z,\lambda)\lambda}{g_k A_{ki}} = C \times \exp\left[-\frac{E_k}{kT_{\mathrm{exc}}(z)}\right] \tag{3.4}$$

where $I(z, \lambda)$ is the measured intensity of the spectral line of wavelength λ, g_k is the statistical weight of the upper state in the transition with probability A_{ki}, E_k is the energy gap in the transition, k is the Boltzmann constant and $T_{\mathrm{exc}}(z)$ is the electron excitation temperature at vertical position z with respect to the air plasma spray baseline (cap surface or electrode surface).

Electron excitation temperature $T_{\mathrm{exc}}(z)$ is then obtained by plotting the normalized spectral line intensity (left side of Eq. (3.4)) as a function of the energy E_k of the transition, and by fitting to the exponential function and evaluating the optimum coefficient in the exponent. The result is presented in Fig. 3.4. As shown, the electron

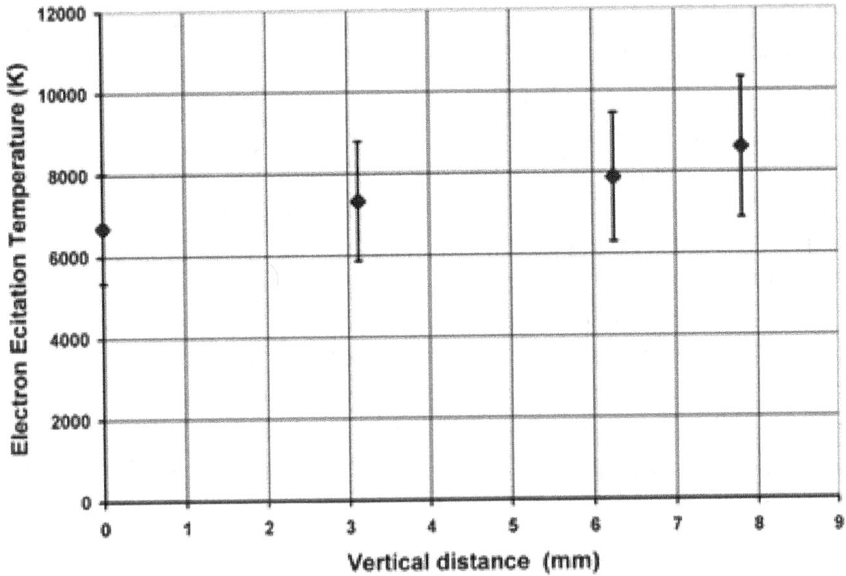

Fig. 3.4.   Axial distribution of the electron excitation temperature.

excitation temperature is about 7700 K, signifying that this air plasma spray is nonequilibrium. Atomic levels used in excitation temperature evaluation had ionization energy of the order of 1–4 eV. One can estimate then that the average electron energy in the non-thermal component of the electron plasma is of the order of several electron volts.

3.6 Impacts of Atomic Oxygen, Ozone, Nitric Oxide, and UV Radiation

Atomic oxygen is very reactive and can oxide most elements. It is essential for respiration of animals and can be used in medical applications. It kills disease microorganisms, improve cellular function, and promote hemostasis and healing of damaged blood vessels and tissues.

The debris from bacterial cells (called endotoxins) are organic but are not living; therefore, it is difficult to remove all from the implants even with modern sterilization practices. These debris can cause inflammation. Atomic oxygen is capable of damaging organic molecules to remove biologically active contaminants from surgical implants. It can clean the implant and remove all traces of organic materials, which greatly diminishes the risk of post-operative inflammation.

Ozone is a powerful oxidant, which inactivates many disease bacteria and viruses. However, it is a toxic gas, harmful to the upper respiratory tract and the lungs.

The nitric oxide molecule is an anti-bacterial agent and provides signaling and regulation biological functions. Human body produces NO to kill invading pathogens. At the same time nitric oxide works as a primary vasoregulator and anti-hypertensive agent. NO also regulates: inflammation, collagen production, angiogenesis and apoptosis. Exogenous NO generated by plasma devices could enhance bio-activity of NO-assisted processes in human organism.

The photon energy in an electromagnetic radiation is inversely proportional to the wavelength of the radiation. In terms of the electron volt (eV) unit, the photon energy ε_P and the wavelength λ is related by

$$\varepsilon_p = \frac{1240}{\lambda} eV \tag{3.5}$$

where λ is in nm and $1 \text{ eV} = 1.6 \times 10^{-19} \text{ J}$.

Ultraviolet (UV) is an electromagnetic radiation with a wavelength from 100 nm to 400 nm, which has the photon energy in the range of 3.1 eV to 12.4 eV. UV radiation is divided into three wavelength ranges: UV-A (315–400 nm), UV-B (280–315 nm), and UV-C (100–280 nm); in which UV-B radiation penetrates uncovered skin to convert cutaneous 7-dehydrocholesterol into previtamin D3, for the vitamin D3 production and UV-C is used in germicidal applications, for example, in the treatment of psoriasis.

On the other hand, UV radiation can cause health hazards. Although UV-C poses the maximum risk, overexposure causes temporary skin redness and harsh eye irritation, but no permanent damage. UV-A is basically harmless. However, prolonged exposure to UV-B can cause skin cancer, skin aging, and cataracts (clouding of the lens of the eye).

Problems

P3.1. An optical spectrometer (often simply called "spectrometer") takes in optical signal (light), breaks it into its spectral components, digitizes the signal as a function of wavelength, and reads the spectral intensity distribution out and displays it through a computer. The first step in this process is to direct the optical signal, e.g., through a fiber optic cable, into the spectrometer through a narrow aperture known as an entrance slit, where the exit signal becomes divergent. The divergent optical signal is then collimated by a concave mirror and directed onto a diffraction grating. The grating then disperses the spectral components of the signal to diffract those at slightly varying angles, which are then focused by a second concave mirror and imaged onto the detector.

A schematic showing the process is presented in Fig. P3.1; because each element in the source leaves its spectral signature in the pattern of lines observed, a spectral analysis can reveal the composition of the object being analyzed.

Once the light is imaged onto the detector the photons are then converted into electric current, which can be calibrated to reveal the spectral intensity. In the case that the source is APS, the spectral intensity around 777.4 nm has a dominant spike in the distribution.

It contains the triplet (777.194, 777.416, and 777.539 nm) of the 5P state OI emission lines. Assume that the emission is isotropic and taking into account that APS has a cylindrical

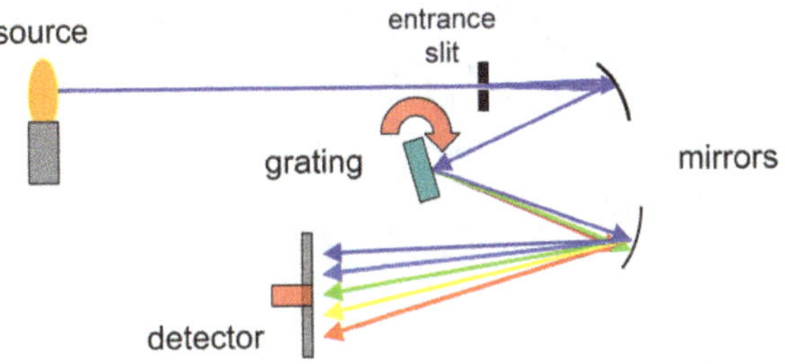

Fig. P3.1. Schematic of the recording process of an optic spectrometer.

shape, the average radiation power density in this spectral region from a slice of the APS at 25 mm away from the nozzle is calibrated to be about P mW/m^2.

(1) Determine the photon flux in Rayleigh.
(2) Estimate the corresponding flux and density of 5P state OI if the flow speed of APS is v m/s.
(3) Let $P = 0.5$ and $v = 20$, calculate the numerical values of (1) and (2).

P3.2. Ultraviolet (UV) light is an electromagnetic radiation with a wavelength from 100 nm to 400 nm, shorter than that of visible light but longer than X-rays, and is invisible to the human eye.

Fig. P3.2 shows the UV band in the optical spectrum. In the range between 200 and 300 nanometers (billionths of a meter), UVs are germicidal, capable of inactivating microorganisms, such as bacteria, viruses and protozoa.

UV disinfection is a physical process, which damages the nucleic acids of microorganisms to inhibit their reproduction and infection functions. Cellular RNA and DNA of microorganisms absorb UV radiation, primarily at 254 nm, to create double bonds and thymine dimers between adjacent

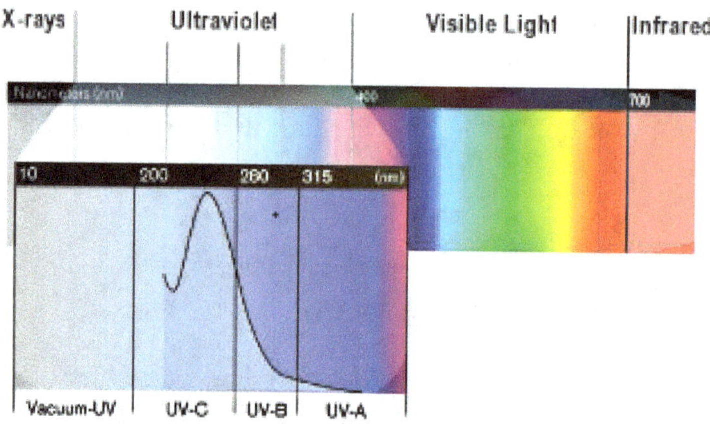

Fig. P3.2.   Optical spectral lines and the spectral ranges of the UV bands.

nucleotides; causing microorganisms' inability to replicate and infect.

UV treatment does not remove microorganisms, it inactivates them but can eventually break down their DNA on a cumulative basis. Thus, UV disinfection systems should deliver sufficient UV dose (e.g., >12 mJ/cm^2 = 1200 μW-sec/cm^2 for most of the microorganisms) to overwhelm DNA repair mechanisms. The amount of dose from the treatment is related to the exposure time and radiation intensity. The exposure time is reported as "microwatt-seconds per square centimeter" (μW-sec/ cm^2).

UV lamps producing 800 μW/ cm^2 germicidal UV radiation of 254 nm at 1m distance are used to disinfect airborne pathogens of a cleanroom. Find

(1) the photon flux of the lamp at 1 meter away; and
(2) estimate the required exposure time if the required total dose is 1 J/ cm^2.

Chapter 4

Plasma Disinfection and Sterilization

4.1 Bacteria

Bacteria grow in colonies and reproduce rapidly by asexual budding or fission, in which the cell increases in size and then splits in two. Bacteria can also undergo conjugation in which two separate bacteria exchange pieces of DNA.

Bacteria are traditionally divided into the two groups: gram-positive and gram-negative, based on their Gram stain retention. Gram-positive bacteria take up the crystal violet stain used in the test, and then appear to be purple-colored when seen through a microscope. The peptidoglycan layer of gram-negative bacteria is much thinner, and is sandwiched between an inner cell membrane and a bacterial outer membrane, causing this group of bacteria to take up the counterstain (safranin or fuchsine) and appear red or pink.

Both gram-positive and gram-negative bacteria commonly have a surface layer called an S-layer. However, there is a distinctive difference between the structures of their cell walls. Gram-positive bacteria are monoderms; they are bounded by a single-unit lipid membrane, and, in general, contain a thick layer (20–80 nm) of peptidoglycan responsible for retaining the Gram stain. On the other hand, gram-negative bacteria are diderms; they are bounded by an inner cytoplasmic membrane and an outer cell membrane, and contain only a thin layer of peptidoglycan (2–3 nm) between these membranes.

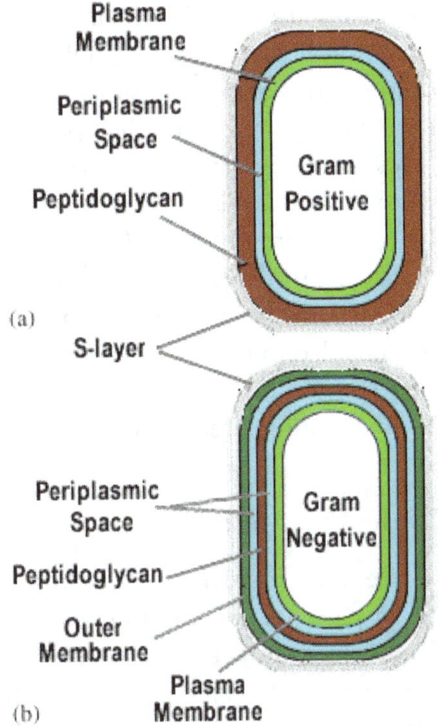

Fig. 4.1. Cell walls outside the cores of (a) gram-positive and (b) gram-negative bacteria.

Two cartoon plots, representing those two kind bacteria, are presented in Fig. 4.1, to illustrate the difference. As shown in Fig. 4.1a, the S-layer of gram-positive bacteria is attached to the peptidoglycan layer. On the other hand, the S-layer of gram-negative bacteria, as shown in Fig. 4.1b, is attached to an outer membrane.

Despite their thicker peptidoglycan layer, gram-positive bacteria are more receptive to antibiotics than gram-negative, due to the absence of the outer membrane. The outer membrane protects the gram-negative bacteria from several antibiotics, dyes, and detergents that would normally damage either the inner membrane or the cell wall (made of peptidoglycan). The outer membrane provides these bacteria with resistance to lysozyme and penicillin.

Gram-positive bacteria are capable of causing serious and sometimes fatal infections in newborn infants as well as dental caries. Oral pathogens are representatives of gram-positive bacteria. If gram-negative bacteria enter the circulatory system, the liposaccharide can cause a toxic reaction. This results in fever, an increased respiratory rate, and low blood pressure. This may lead to a life-threatening condition of septic shock. *E. coli* is a representative of gram-negative bacteria.

Under unfavorable environmental conditions, some of bacteria, called spores, develop a thick outer wall and enter a dormant phase. Bacterial spores then evolve to endospore form and remain in dormant state, able to survive for long periods of time under the conditions that kill many other organisms.

4.2 Issues of Disinfection Methods

According to the Centers for Disease Control and Prevention (CDC), 8% to 10% of patients become infected while in the hospital; hospital-acquired infections (HAIs) kill nearly 100,000 people a year. To address the concerns over HAIs, considerable attention is paid to the sterilization of medical tools and devices, critical to patient safety. An "ideal" sterilization method is highly efficacious, rapidly active; has strong penetrability, materials compatibility; is non-toxic, organic material resistant; has adaptability, monitoring capability, and is cost-effective.

Disinfection and Sterilization of the healthcare facilities include three classifications:

1. Processing "Critical" patient care objects such as surgical instruments and devices; cardiac catheters; implants; etc.

   Those objects normally penetrate the sterile tissue, or the vascular system of patients, through which blood flows; thus those have to be sterile, and all microorganisms including bacterial spores, have to be eliminated.

2. Processing "Semicritical" patient care objects such as respiratory therapy and anesthesia equipment, Gastrointestinal (Gi) endoscopes, endocavitary probes, tonometers, diaphragm fitting rings, etc.

Those Objects come in contact with patient's mucous membranes or skin that is not intact; those require high level disinfection which kills all microorganisms except high numbers of bacterial spores.

3. Processing "Noncritical" patient care objects such as bedpans; crutches; bed rails; ekg leads; bedside tables; walls, floors and furniture.

Those objects will not come in contact with patient's mucous membranes or skin that is not intact; thus only vegetative bacteria, fungi and lipid viruses need to be killed.

The standard disinfection and sterilization methods are using pressurized hot air of 170°C, pressurized hot water vapor of 120°C, and wet chemistry. To help cut the number of HAIs, hospitals are also using highly aggressive disinfectants, including alcohol, chlorine and chlorine compounds, formaldehyde, hydrogen peroxide, and ortho phthalaldehyde, and overkill sterilization methods, such as autoclave, e-beam and gamma radiation, and ethylene oxide (EtO), which can cause plastic medical devices to crack, lose critical properties, or change color.

However, the selection of the sterilizing agent to achieve sterility depends primarily upon the nature of the item to be sterilized. The selected agent has to be able to contact the all surfaces of the item, and the time required to complete the process of killing spores in the equipment thoroughly is critical. Although none of the physical or chemical processes can absolutely destroy all pathogens and microorganisms, supplies and equipment are yet considered sterile when necessary conditions have been met during a sterilization process.

To ensure that instruments and supplies are sterile when used, monitoring of the sterilization process is essential. It uses mechanical indicators to monitor functions of each sterilizer, as appropriate; uses a chemical indicator on a package, which can verify exposure to a sterilization process, and can detect sterilizer malfunction or human error in packaging or loading the sterilizer; and uses a biologic indicator to detect non-sterilizing conditions in the sterilizer. A biologic indicator is a preparation of living spores resistant to the sterilizing agent. These may be supplied in a self-contained system, in dry spore strips or discs in envelopes, or sealed vials or ampoules of spores to be sterilized and a control that is not sterilized; some also incorporate a chemical indicator.

As hospitals continue to increase the number and diverse types of expensive, intricate instruments, their sterilization demands quicker turnaround times; and the overkill sterilization methods become less favorable.

Residual tissue contamination on surgical instruments is also a growing concern in the healthcare industries. The proteinaceous debris following microbial death during sterilization has to be removed from device surfaces to minimize the risk of inflammatory responses of implantable medical devices. The body's immune system responding to non-infectious contamination can cause a severe inflammation that can lead to cell death.

Applying traditional hospital decontamination and sterilization methods to a range of "in-use" stainless steel surgical instruments, after several cleaning cycles protein contaminants were still detected. On the other hand, no retention of contamination was evident on the surfaces of instruments that were subjected to gas-plasma cleaning.

Thus, new sterilization methods using CAPs as disinfectants are being explored. The salient features of CAP disinfectant include:

1) It is dry gas and does not produce chemical wastes;
2) it does not pose chemical corrosive effect;
3) it is fast working and offers ultra-fine cleaning, leaving no residues;
4) it is low temperature, without harming the medical tools and instruments;
5) it does not require mass storage, and in the case of air plasma, as ambient air is the working gas, no gas tank is needed;
6) it is adaptable for sensitive surfaces, including living tissues and open wounds, in dermatological treatment.

Atmospheric pressure plasma converts electromagnetic (EM) energy to the kinetic energies of electrons and ions, to the internal energies of molecules and atoms, and to thermal energy. Non-equilibrium plasma makes a better usage of the applied EM energy for disinfection/sterilization applications by converting less EM energy into heat; it also minimizes the thermal damage in the treatment. The atoms and excited molecules become reactive and some of those emit ultraviolet (UV) radiation.

The main sterilization process is decomposition of organic molecules of living organisms by bombardment with electrons, ions and short-lived neutral reactive species carried in plasma, in which OI atoms are capable in the cleavage of organic molecular chain by abstracting hydrogen from the molecules. When reactive oxygen species and atomic oxygen interact with living organisms, radicals and radical sites in the organic molecules are produced. Cleavage of radical organic molecules makes them easy to be oxidized in air. In addition, the produced hydroxyl radicals (*OH) react quickly with organic molecules to produce alkyl radicals, which also rapidly oxidize in air. With this oxidation mechanism, reactive oxygen species (ROS) and atomic oxygen indiscriminately attack all organic molecules of living organisms, including those in bacteria capsules and cell walls. When the rate of the cell wall damage is higher than the rate of its own reparation, it leads to the death of the cell. Hydrogen peroxide (H_2O_2) is less reactive than the *OH radical and atomic oxygen OI. On the other hand, it has a better chance of penetrating into the bacterial cell nucleus, where it can damage the DNA to cause the death of the cell.

Tests in vitro on different microorganisms have been performed to explore the efficacy of plasma treatment on microbial inactivation. Many kinds of microorganisms are protected by a layer of biofilm, which makes it difficult for conventional disinfectants and even antibiotics to kill them. On the other hand, reactive oxygen species (ROS) kill the microorganisms because those species can penetrate through the biofilm as well as the membrane of bacteria.

4.3 Plasma Disinfection Approaches

Non-thermal plasma for medical application was classified into 'direct' and 'indirect' approaches. The direct approach, such as FE-DBD, applies a dielectric-covered high voltage electrode to the treated tissue/skin which acts as the counter (floating) electrode to setup discharges directly on the surface of treated tissue/skin. ROS and UV radiation generated in the plasma are effectual in therapeutic applications, such as therapy of skin diseases, skin disinfection, blood coagulation, and enhancing the wound healing process.

Staphylococci, Streptococci and Candida species of yeast in skin samples were treated by FE-DBD for 15 s. A reduction of colony forming units per milliliter from an initial value of 10^9 before treatment to a final value of 10^4 was observed. UV radiation is considered as one of the most important factors in this plasma sterilization. FE-DBD treatment kills melanoma cancer cells through necrosis in 15 s at a plasma power of 14 mW/mm^2; while at low doses (8 mW/mm^2) necrosis was absent, but after about 5 sec treatment, apoptosis was observed later.

In the indirect approach, active species is delivered by the plasma, injected from a plasma generator, to the treated surface. The bacterial inactivation efficacy of the plasma pencil was tested using *E. coli* in agar. Plasma was generated in helium with 0.75% O_2 as admixture. After two minutes treatment, such an admixture helium plasma creates a zone of inactivation in agar which has a diameter almost double that created by a pure helium plasma. The inactivation efficacy of the plasma was improved significantly with the admixture of O_2, which supplied resource to generate oxygen-based reactive species. It was concluded that oxygen-based reactive species play an important role in inactivation of the bacteria.

Most of the active species are lost during transport; it is due to chemical and physical processes in the effluent, like electron-ion recombination and combination of atoms into stable molecules. Thus DBDs are gaining importance since they can ignite plasma directly on the body surface which is at floating potential. However, many of the disinfection applications favor the use of the indirect approach, which is more applicable to disinfect areas where a direct use device is unable to make contact. On the other hand, an air plasma spray (jet) can generate much abundant oxygen-based reactive species to compensate their loss during the transport. Air plasma spray exploits the bactericidal and virucidal effect of ROS for disinfection; the test results show that such biochemical mechanism achieves an even higher bacteria kill rate than that of a direct use device (e.g., FE-DBD) and have no apparent side effects.

CAP also offers healthcare which includes treatment of cancer cells and the initiation of apoptosis, prion inactivation, prevention of nosocomial infections, and the therapy of infected wounds.

4.4 CAPs on Dental Issues

Dental caries are the localized destruction of tooth tissue by the acids produced by bacteria. Caries start with small demineralization areas under the enamel, which progress through the dentine to the pulp. *Streptococcus mutans*, a gram-positive bacteria, is one of the major causes of caries. Periodontal disease is related to dental plaque, which is a complex oral biofilm formed with several microbial species. It infects the gum, causing it to detach from the tooth. Root-canal treatment may fail if, for instance, *Enterococcus faecalis* bacteria are not disinfected thoroughly.

Disinfectants, such as liquid antibiotics, sometimes fail to kill all the bacteria living deep inside the sick tooth. This is because gram positive bacteria of *S. mutans* tend to group together to form biofilms, which are able to tolerate antimicrobial challenges from antibiotics.

There is considerable progress on exploring the applications of CAPs for dental disinfection, biofilm removal and prevention, instrument sterilization, composite restoration, and dental bleaching. When CAP is applied for dental treatment in the mouth, only the indirect approach by a plasma jet is feasible. The antimicrobial effects of ROS in the plasma effluent is a means to eradicate oral pathogens, remove dental biofilms, and disinfect root canals.

In the next sections, tests of air plasma treatments on gram positive bacteria and biofilms, which have the strongest resistance to the conventional disinfection, are described and discussed. An air plasma spray (APS) was used in the tests. The efficacy of the treatment on oral pathogens and dental biofilms were studied.

4.5 Experiments

A defined microorganism culture suspended in the sterile sodium chloride solution (0.89% of NaCl solution) was inoculated in an agar plate as a test sample. The concentration of microorganisms in the cultivating medium was approximately 10^6 colony forming units (CFU)/mℓ to form continuous overlay of microorganism coats on the agar plate. Samples were then exposed to the plasma effluent for different treatment times. There are two types of treatment approaches, 1) continuous approach which keeps the plasma generator running for the entire treatment period, and 2) intermittent approach which turns the plasma generator on and off periodically until the accumulated turned-on time reaches the desired treatment time. After the plasma treatment, samples of different microorganisms and different treatment times were cultivated at the best condition for each microorganism to observe the dependency of the zone of growth inhibition on the total treatment time. Results were compared to comprehend the effect of plasma treatment on different microorganisms. The surface temperature of the sample holder, which usually increases with the exposure time, was monitored.

4.6 Air Plasma Treatment of Oral Pathogens

Included in the investigation were six oral microorganisms, *Actinomyces naeslundii* CDCA1916 (ATCC12104), *Candida albicans* NIH3172 (ATCC14053), *Streptococcus gordonii* NCTC7865 (ATCC10558), *Streptococcus mutans* UA159 (ATCC700610), *Streptococcus oralis* NCTC11427 (ATCC35037) and *Streptococcus sanguinis* DSS-10 (ATCC10556). Each test was repeated three times.

4.6.1 *Experimental conditions*

The emission spectroscopy of an air plasma spray (APS), similar to that shown in Fig. 2.5a and run by the same power supply shown in Fig. 2.6a, was examined. Spatial distributions of the emission intensities of the OI lines at 202.64 nm, 770.68 nm, and 777.19 nm are presented in Fig. 4.2.

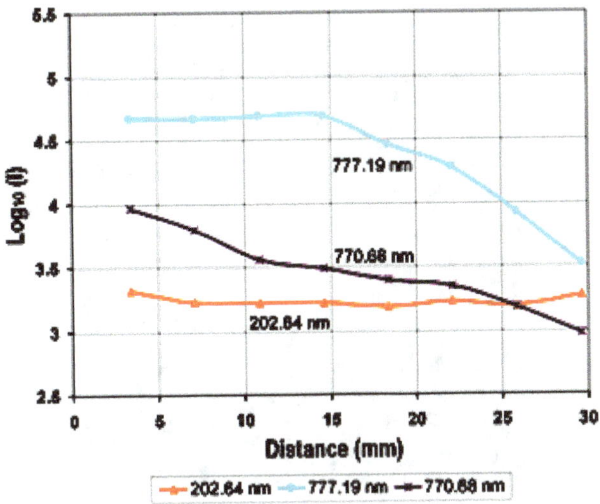

Fig. 4.2. Spatial distributions of the emission intensities of the OI lines at 202.64 nm, 770.68 nm, and 777.19 nm.

It shows that the intensity at 777.19 nm is the strongest. The intensities of these lines drop rapidly as one move away from the nozzle exit of the generator; nevertheless, those line intensities at 30 mm away still remain at significant levels. Nitric oxide (NO) bands with band heads at 237.02 and 247.87 nm were also measured. It indicates that APS also delivers low level NO flux in the treatment.

This air plasma spray was secured on a fixed vertical holder with its circular nozzle exit facing downward to a sample holder, which is made of agar placed inside a glass dish, and aligned at a distance of 30 mm below the nozzle exit, as demonstrated in Fig. 4.3.

Samples were treated by this air plasma spray (APS), applying an intermittent approach which runs APS with 5-second on/10-second off in one cycle. Four different numbers of cycles, 1, 2, 4 and 6 at a fixed exposure distance of 30 mm were performed. The tests produced four sets of results from four different accumulated treatment times of 5, 10, 20, and 30 s for comparison.

The surface temperature of the sample holder, made of agar placed inside a glass dish, increased with the exposure time; however,

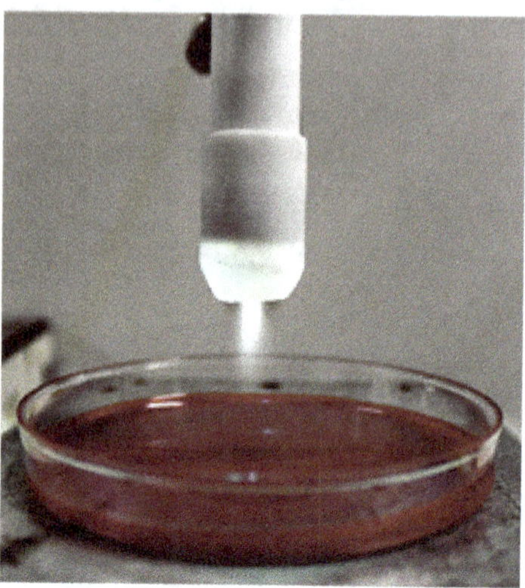

Fig. 4.3. A photo of the experimental setup.

the intermittent approach effectively limited temperature elevation, which was up from room temperature (~ 23°C) to about 37°C after 30 seconds accumulated exposure time (though the temperature of the APS was measured to be around 47°C). Thus the thermal effect on the samples was minimized.

4.6.2 *Zone of inhibition of microorganisms by plasma treatment*

After cultivation of the treated sample, the central region, where was exposed to the plasma effluent, does not grow colony forming units (CFU) of the microorganisms and thus is transparent to the agar plate; whereas in each of the controls, microorganisms have grown everywhere in the surface, which becomes opaque to the agar plate. This is shown in Fig. 4.4 by the images of the (a) control and (b) treated *S. mutans* samples.

A zone of growth inhibition is circled in Fig. 4.4b of the treated sample, and the zone of inhibition is defined by the diameter (mm) of the circle. It was found that there is a dose-response relationship showing an increase of the diameter of the zone of inhibition with an increase of the exposure time.

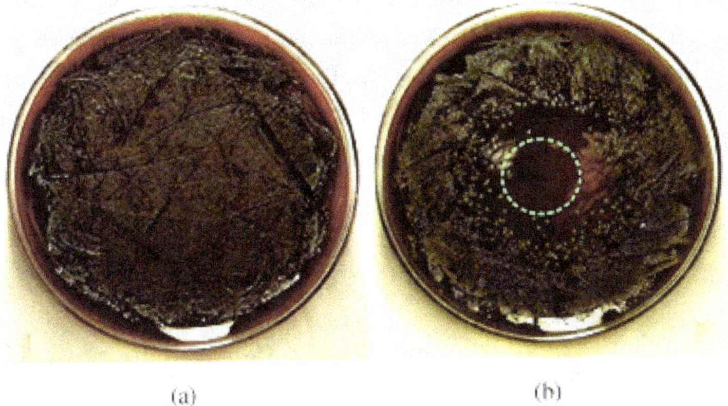

Fig. 4.4. Comparison of the growth situations between (a) the control and (b) the treated sample. The circle in (b) spots the zone of growth inhibition.

Table 4.1 Diameter (mm) of the zone of inhibition versus the plasma treatment time.

	Treatment				
Microorganism	Plasma 5 s	Plasma 10 s	Plasma 20 s	Plasma 30 s	Air flow (control) 30 s
Actinomyces naeslundii CDCA 1916	0	20	25	30	0
Candida albicans NIH 3172	0	8	17	21	0
Streptococcus gordonii NCTC 7865	25	27	30	35	0
Streptococcus mutans UA 159	20	27	32	35	0
Streptococcus oralis NCTC 11427	0	25	30	35	0
Streptococcus sanguinis DSS-10	26	27	30	30	0

Samples of six microorganisms were treated by APS with the periodic approach at four different total exposure times: 5 s, 10 s, 20 s and 30 s; each test was repeated three times. The average diameters of measured zones of growth inhibition are summarized in Table 4.1.

The results, summarized in Table 4.1, show that the APS creates a zone of microbial growth inhibition in each treated sample and the diameter of the zone of inhibition increases with the plasma treatment time. In fact, after 20 s treatment, zones of microbial growth inhibition in all samples expand to be larger than the entire region directly exposed to the APS. These observations demonstrate that APS is effectual to disinfect the gram-positive bacteria and fungi, it also attracts to microorganisms. Atomic oxygen carried by the APS is the bactericidal agent.

4.7 Dental Disinfection

The disinfectants, such as liquid antibiotics, sometimes failed to kill all the bacteria living deep inside the sick tooth. This is because dental bacteria tend to group together to form biofilms as dental plaque, where they may cause tooth decay and gum disease.

Biofilms are densely packed microorganisms in which cells are embedded within a self-produced matrix of extracellular polymeric substances (EPSs) which adhere to each other and/or to a surface. EPSs form a barrier to block the penetration of antibiotics and chemicals. Cells deep within a biofilm grow fairly slowly and so are less susceptible to antibiotics. Furthermore, biofilms contain zombie-like "persister" cells which lie dormant when antibiotics are present but spring into action after antibiotic treatment ends. Cells within biofilms also organize themselves to pump drugs right out of cells via "a kind of bulimic behavior". These explain why biofilms are able to tolerate antimicrobial challenges from antibiotics.

In the following, study of cold air plasma as a new type of dental disinfectant is given. The procedure of preparing biofilm samples for in vitro tests is first described. The tests on preventing the formation of biofilms (case A) as well as on killing mature biofilms (case B) by cold air plasma are then presented. The results demonstrating the efficacy of cold air plasma in sterilizing biofilm are discussed.

4.7.1 *Experiment preparation and procedure*

S. mutans UA159 were inoculated in the Tryptone-yeast extract broth containing 1% sucrose to form biofilms on saliva-coated hydroxyapatite discs placed in a vertical position (HAP ceramic — calcium hydroxyapatite, 0.5″ diameter — Clarkson Calcium Phosphates, Williamsport, PA) in batch culture at 37°C and 10% CO_2. Normally, it takes five days of continuous cultivation of the sample with the Tryptone-yeast extract broth and 1% sucrose, where the culture medium is changed daily, to establish a mature biofilm (which has a thickness of about 200 µm) on a saliva-coated hydroxyapatite disc. Thus, all the biofilm samples employed in the study received five day cultivation. Each treatment was 30 s continuous exposure to APS at 30 mm distance from the nozzle outlet of the plasma spray. In the control group, similar biofilms were treated using the same airflow from the unignited plasma spray also for 30 s in each treatment. The treatment time used in the experiment was based on the zone of inhibition data, presented in Table 4.1, indicating thorough disinfection in 30 s.

4.7.1.1 *Plasma treatment effect on biofilm formation*

After 24 h of initial biofilm formation, the biofilms were treated twice daily until day 5; each treatment is 30 s continuous exposure at 30 mm distance. A flow chart with the treatment plan of case A is presented in Fig. 4.5. In this study, the total number of 30 s continuous plasma treatments was eight times.

4.7.1.2 *Plasma treatment effect on biofilm disinfection*

In case B, a mature biofilm was treated once to explore the disinfection capacity of the plasma spray.

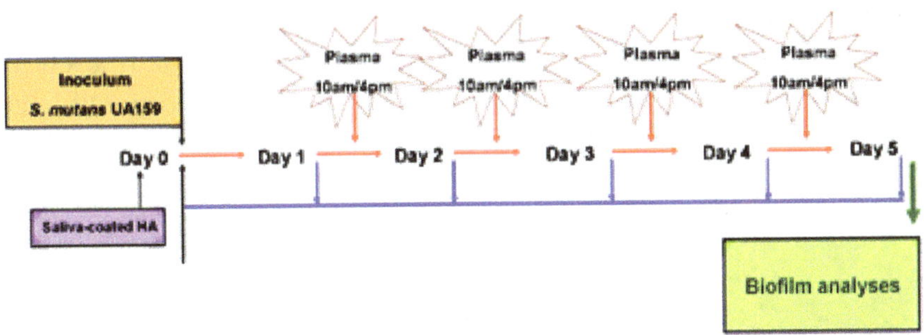

Fig. 4.5. A flow chart for biofilm formation tests.

After the treatments in both cases, the biofilms (n = 12) were harvested for the analyses of 1) bacterial viability, 2) dry-weight, 3) polysaccharide composition content by colorimetric assays, 4) morphology, using environmental scanning electron microscopy (ESEM), and 5) fluorescence microscopy was also used to show live and dead bacteria after plasma treatment.

4.7.2 Experimental results

The biofilm formed on the surface of a saliva-coated hydroxyapatite disc becomes mature after 5 days of unperturbed formation. In this state, aggregations of microbial cells are bound together by an extracellular matrix of polysaccharide and protein and the microorganisms become able to tolerate antimicrobial challenges that normally eradicate free-floating individual cells.

In the experiments, plasma treatment was introduced to A) prevent the biofilm formation and B) disinfect mature biofilm. The treated biofilms and the corresponding controls were analyzed to compare their bacterial viability, biomass, and accumulated amounts of polysaccharide.

4.7.2.1 *On preventing biofilm formation (case A)*

Figure 4.6 shows that the dry-weight of the biomass and the amount of insoluble polysaccharide were reduced by more than 80%. The results indicate that the day 2 treatments (the first two of the 8 treatments) could already inhibit the growth of CFU and slow down the biofilm forming considerably. The 8 orders of magnitude reduction in viable CFU demonstrates that the daily treatment with APS can fully inhibit the biofilm formation on the saliva-coated hydroxyapatite disc.

4.7.2.2 *On biofilm disinfection (case B)*

In this case, the viable CFU was also reduced by about 8 orders of magnitude; however, contrary to the case A, there were only small

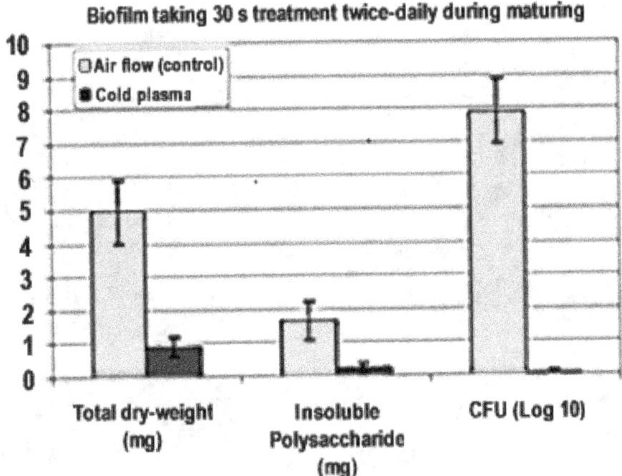

Fig. 4.6. Changes of the composition and viability of the treated biofilms of *Streptococcus mutans* UA159 in reference to the controls; APS treatment during the formation.

reductions on the biomass and the amount of polysaccharide. This is because the biofilm was already mature. The dry-weight is mainly due to the dead bodies of the microorganisms. The 8 orders of magnitude reduction in viable CFU evidences that APS can disrupt the extracellular matrix of polysaccharide to effectively kill the microorganisms.

4.7.2.3 *Microscope observation*

An environmental scanning electron microscope (ESEM/EVO 50 Zeiss, Germany) was used to examining the morphologies of the treated biofilm and its control in case A (concerning biofilm formation), where the control was treated in a similar fashion, except that the discharge was off (i.e., with airflow on only). There was no need of any pre-preparation in the specimens, the biofilms were analyzed directly in the microscope. The ESEM microscope is capable of imaging delicate hydrated specimens in high pressure, by maintaining a humid atmosphere to prevent water loss in the specimens. The ESEM images of the control and the plasma treated biofilms are presented in Figs. 4.7a and b for comparison.

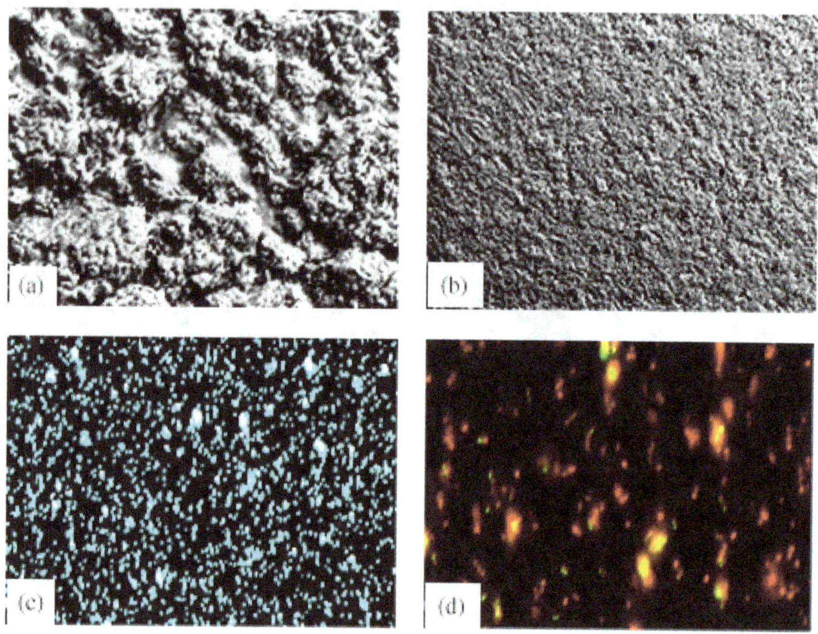

Fig. 4.7.   The architecture and surface topography of biofilms after 5 days of formation examined by environmental scanning electron microscope: (a) biofilm treated with air flow twice-daily during the formation as a control; (b) biofilm treated with the plasma effluent twice-daily during the formation. Fluorescence microscopy stained with **LIVE/DEAD** backlight bacterial viability: (c) mature biofilm treated with 30 s air flow as a control and (d) mature biofilm disinfected by APS for 30 s (Red: dead cells; Green: live cells).

There are no polysaccharide structures, and no biomass distribution can be seen in the image of Fig. 4.7b. It indicates that biofilm formation was fully inhibited with the daily plasma treatment.

The bacterial viability of the control and treatment biofilm in case B was examined via fluorescence microscopy. The live/dead backlight bacterial viability kit (Molecular Probes, Inc. Oregon, USA) was used to stain the live and dead bacteria with green/red colors, respectively. The results are presented in Figs. 4.7c and d. As shown, the viable CFU spreads out in the control. On the other hand, in the treated biofilm, no viable CFU is visible and the dead cells are accumulated in local regions.

4.8 Summary

Experiments of plasma treatments on six defined microorganism cultures as well as on biofilm formation were conducted. The results, summarized in Table 4.1, show that the zone of microbial growth inhibition increases with the plasma treatment time and expands to exceed the cover size of the APS after 20 s treatment. The bactericidal effect of the APS is verified.

The plasma treatments during 5 days of biofilm formation did affect the biomass quantity and the polysaccharides composition, and the viability of the microorganisms in the biofilm, as shown in Fig. 4.6. The structural images, presented in Figs. 4.7a and b, indicate that plasma treatment fully inhibited the biofilm formation. The polysaccharide structures can hardly be seen in Fig. 4.7b.

The experimental results also show that the APS can in fact eradicate the microorganisms deep in a mature biofilm (5 day-old), which usually cannot be eradicated by the traditional antimicrobial agents. The viable microorganisms in the treated mature biofilm sample also drops to the insignificant level. This is likely, because the APS disrupts the polysaccharides from the biofilm matrix; this exposes and attracts microorganisms to the plasma effluent, increasing the biocidal effect. Contrary to that shown in Fig. 4.6 of the prevention case, there were only small reductions on the biomass and the amount of polysaccharide. This is because the biofilm was already mature. The dry-weight is mainly due to the dead bodies of the microorganisms.

Atomic oxygen (OI) carried in APS is an effective oxidant. The oxidation process causes enzymatic degradation of extracellular DNA, which can weaken the biofilm structure and release microbial cells from the surface. Nitric oxide (NO) can further trigger the dispersal of biofilms. The dispersed cells become vulnerable.

4.9 Discussion

Hospital-acquired infections (HAIs) are a major challenge for healthcare professionals and patients, and have gained an increasing importance for overall public health. Every year, HAIs affect more than a million people, and increase mortality, morbidity rates and healthcare costs. Moreover, holding of escalators, touch buttons of elevators, handles of doors, etc., in public areas are also sources of infections during the breakout of various viral diseases. Commercially available biocides used in medical practice are not sufficient to sterilize sources of nosocomial and public infections. Furthermore, antibiotics remain ineffective against multidrug resistant virulent pathogens. Therefore, a new non-toxic, broad-spectrum, dry antimicrobial agent is needed.

CAPs, in particular for CAAP, have shown success in sterilization of a wide range of microorganisms including bacteria, fungi and algae and has even shown success in killing bacterial biofilms. This is because plasma discharges with a high oxygen concentration in CAAP generate an abundance of atomic oxygen (OI), as well as other reactive oxygen species (ROS), such as nitric oxide (NO) and ozone (O_3), which introduce oxidative damage to rapidly inhibit microbial survival as well as to weaken the biofilm structure. Nitric oxide has the potential for the treatment of patients that suffer from chronic infections caused by biofilms.

The generators of CAAP can be designed to be handheld, battery powered, and portable. These factors have brought this exciting new, emerging technology to the forefront of novel antimicrobial techniques. The research areas include the development of CAAP generators and their applications; further studies of the sterilization of bio-medical materials and devices, of the treatment of skin diseases and dental caries, as well as of the healing of chronic wounds, and accelerating wound closure in chronic lesions. Furthermore, the use of CAPs in tumor treatment is a promising and currently intensively researched field as discussed in the next section.

4.10 Cancer Treatment

There is extensive research in developing CAP as a new cancer treatment technology. In contrast to most anti-cancer methods such as chemotherapy and radiotherapy, CAP showed selective inhibition of cell proliferation of cancer cells and normal cells. It triggered more apoptosis in cancer cells than in normal cells. CAP also caused different changes in gene expression. The differential expression of the intracellular anti-oxidant enzymes may be a plausible mechanism to control the selective diffusion of reactive species across the cytoplasmic membrane of cancer cells and selective rise of intracellular ROS in cancer cells.

Due to stronger metabolism in cancer cells, normally the basal ROS level in cancer cells is higher than that in normal cells. When additional ROS stress such as the CAP-originated reactive species is exerted on cells, the whole intracellular ROS in cancer cells will pass a threshold more easily than that in normal cells. A significant rise of intracellular ROS causes intense DNA double-strand breaking; as a result, cancer cells experience more apoptosis than normal cells upon CAP treatment.

The selectivity of plasma effects on cells gives higher susceptibility to the cancer cells in the CAP treatment. Preliminary research over the past decade demonstrated that CAP could effectively inhibit the growth of dozens of cancer cell lines in vitro by mainly triggering apoptosis. Applying treatments on cultured human melanoma and hepatocellular carcinoma cells, it is able to induce several modes of malignant cell death, including early or late apoptosis and necrosis, which are caused by DNA fragmentation and the release of the mitochondrial protein cytochrome. CAP is also capable of effectively resisting the growth of subcutaneously implanted xenograft tumors in mice. In sum, the results observed on in vitro and a few in vivo studies show that CAPs induce cell growth arrest and apoptosis; and reduce cell migration and invasion activities. The decrease of metabolic activity, the induced apoptosis, and the reduction of viable cancer cells are in direct proportion to the duration of CAP treatment.

Local CAP treatment at a tumor site may enhance drug delivery efficacy, advancing chemotherapy to targeted therapy. A test indicates that CAP treatment could restore the sensitivity of chemo-resistant glioma cells. 60 seconds of CAP treatment combined with 100/200 mM therapy with temozolomide (TMZ) showed a statistically significance increase in inducing a cell-division cycle arrest compared with TMZ treatment alone.

Although the anti-cancer molecular mechanism is still far from clear, CAP shows its promising potential to be an anti-cancer tool. More work is required to make CAP useful in the clinic. Identifying a therapy for cancer is challenging because the treatment has to preferentially attack the cancer cells and let the normal cells live. The mechanism that the cancer cells are more sensitive to CAP treatment than normal cells has to be further verified and understood in order to paving CAP as an ideal cancer therapy. The key questions in further R & D of the technology are:

1. What is the advantage of CAP compared with other existing treatment modalities?
2. What is the essence of the selective anti-cancer capacity of CAP?
3. What are the distinct cellular and molecular responses between cancer cells and normal cells to the CAP treatment?
4. What are the general cellular and molecular features or markers to describe such killing selectivity?
5. Can such a selectivity be enhanced by adjusting CAP generation?
6. Which are the cancer subtypes more sensitive to the CAP treatment?
7. How can the inhibited growth of subcutaneous tumors be understood based on the results of CAP treatments just above the skin?
8. What is the potential toxicity of CAP?
9. Can the toxicity be eliminated by adjusting CAP generation?

Problems

P4.1. Disinfectants are antimicrobial agents that are applied to the surface of non-living objects to destroy the cell wall or to interfere with the metabolism of microbes living on the objects.

Phenol is the standard to be compared with other disinfectants for the effectiveness on a standard microbe (usually *Salmonella typhi* or *Staphylococcus aureus*), and the corresponding rating system is called the "Phenol coefficient". Disinfectants that are more effective than phenol have a coefficient > 1. Those that are less effective have a coefficient < 1.

Set up tests to check the Phenol coefficients of hydrogen peroxide and alcohol.

P4.2. Bacteria, also called germs, are single-celled, or microscopic organisms. Though small, bacteria are powerful and complex, and they can survive in extreme conditions. Bacteria have a tough protective coating that boosts their resistance to white blood cells in the body. Some bacteria are good and some are bad. Harmful bacteria are called pathogenic bacteria. List the useful benefits of good bacteria.

P4.3. Viruses may be classified in three groups: 1) Group A: lipid-containing viruses (e.g. *Herpesviridae*, *Paramyxoviridae*, *Orthomyxoviridae*); 2) Group B: small (20–30 ran), non-lipid viruses (e.g. *Picornaviridae*, *Parvoviridae*); and 3) Group C: other non-lipid viruses (e.g. *Adenoviridae*, *Reoviridae*, *Papovaviridae*).

In vitro tests show that group A viruses are uniformly susceptible to all disinfectants to a high degree; but viruses without lipid and small size (i.e., Group B and C) are resistant to lipophilic chemical agents. On the other hand, oxidizing agents are active against most viruses.

Air plasma carries reactive oxygen species (ROS). Explain the virucidal mechanism of air plasma disinfection.

P4.4. The CAP's virucidal process progresses in virus life cycle. By adjusting the plasma treatment dose (e.g., the CAP treatment time), the impact of CAP on the virus activations in different stages of virus life cycle may be explored. The results help to reveal the CAP virucidal mechanism. What are the virus activations in different stages of virus life cycle?

P4.5. Reactive species play an important role in the sterilizing mechanism of plasma. It has been shown that reactive oxygen species (ROS) generated in discharges have a strong germicidal effect. In the use of gas plasma in inactivation of bacteria, the inactivation rate varies with the type of the bacteria and the type of working gas. It is observed that gram positive bacteria are more sensitive to plasma treatment than gram negative bacteria; helium plasma is hardly effective against bacteria; argon plasma is better, but the best results are obtained from air plasma. Based on likely germicidal mechanisms, explain the observations.

P4.6.   Plaque is a biofilm on the surfaces of the teeth. These densely packed microorganisms are embedded within a self-produced matrix of extracellular polymeric substances (EPSs) and subject the teeth and gingival tissues to high concentrations of bacterial metabolites which results in dental disease. Explain why enzymatic degradation of extracellular DNA can weaken the biofilm structure and APS can be so effective to kill biofilm, evidenced by the results presented in Fig. 4.7d.

P4.7.   Highly efficient fluorescent proteins such as the green fluorescent protein (GFP) have been developed using the molecular biology technique of gene fusion, a process that links the expression of the fluorescent compound to that of the target protein. Genetically modified cells or organisms directly express the fluorescently tagged proteins, which enables the study of the function of the original protein *in vivo*. Many different fluorescent dyes can be used to stain different structures or chemical compounds of cells. When those are illuminated with high energy (excitation) light, they emit lights at designated colors. An ideal fluorescent image shows only the structure of interest that was labeled with the specific fluorescent dye. On the other hand, different fluorescent dyes can be used together to stain different biological structures, which can then be detected simultaneously, while still being specific due to the individual color of the dye. Ultraviolet at 280 nm is used as the excitation light, what is the photon energy in eV?

P4.8.   What are the three types of indicators used to monitor sterilization parameters?

P4.9.   What would be a specific goal in a high level disinfection (HLD) or Sterilization Risk Assessment?

Chapter 5

Plasma Decontamination

5.1 Background

Decontamination is an action to remove hazardous materials from contaminated areas. The contaminants include chemical, biological, radiological, and nuclear (CBRN) agents. The objective of decontamination is to clean reusable equipment, instruments, and supplies in the decontamination area and protect the preparation and package workers who come in contact with those devices in the decontamination process from contracting diseases caused by microorganisms on those devices. Therefore, awareness of the decontamination procedure is important. The steps in the decontamination process may be listed as

1. Personnel working in the decontamination process wear protective attire.
2. Contaminated objects are covered properly to avoid accidental contamination before being transported to the decontamination area.
3. Contaminated objects are sorted to remove disposable sharps, discard other single-use items, and soak those items filled with debris or covered with blood, before proceeding to decontamination.
4. Decontamination is processed with chosen method(s) and decontaminant(s).

5. All decontaminated items should undergo inspection to make
    sure they are sterilized, and that there are no damages or mal-
    function before being packaged, for reuse or storage.

Biological emergencies are situations where microorganisms, which are difficult to be treated, are released unexpectedly to cause illnesses and death. Biological warfare agents (BWA) consist of viruses (e.g., Ebola, Yellow Fever, Severe Acute Respiratory Syndrome (SARS), Smallpox), spore forming bacteria (e.g., Anthrax), vegetative bacteria (e.g., *Escherichia (E.) coli*, Plaque), and biotoxins (e.g., Botox, Ricin). Due to the highly fatal nature of pulmonary anthrax (80–90%), the ease of production and storage of the spores of that *Bacillus* and their survival in the environment after bioattack, this organism (Anthrax) has become the primary bacterial agent in bio-warfare and bioterror attack. For instance, their use in the terror attacks of 2001, via mails, has brought the issues surrounding the deliberate release of BWA into sharp focus. To counter the threat of terrorist attacks, an effective decontamination defense is required to minimize the consequences of biological attacks. Currently, different methods can be used to decontaminate property and persons exposed to biological contaminants.

Biological decontamination services are also frequently required for a wide range of healthcare, pharmaceutical and research facilities where high levels of cleanliness are required to prevent a risk to health through exposure to biological pathogens.

The traditional decontamination methods for BWA involve the use of "wet" solutions, which include bleaches and Decontamination Solution #2 (DS2). These hazardous chemicals need to meet special guidelines for storage, transport, and disposal during and after usage. Many of the toxic chemicals involve delays while chemicals disperse to safe levels, which could present a risk to decontamination personnel. Moreover, "green" decontaminants are preferred because these chemicals are released into the environment. The decontamination time of these methods is typically 30 minutes. To reduce the exposure time, Sandia National Laboratories have used a cocktail of ordinary substances to create a type of foam, which can neutralize viral, bacterial, and nerve agents in minutes, as reported in the internal news release on March 1, 1999. The main active ingredient of the formulation is hydrogen peroxide. How the foam kills spores is not well understood. It is thought the surfactants perforate the spore's protein armor and allow the oxidizing agents (mainly, H_2O_2) to attack the genetic material inside. Tests conducted at Sandia showed that the foam destroyed simulants of VX, mustard, Soman, and anthrax.

Decon Green is a decontamination solution developed by the U.S. Army to detoxify chemical and biological warfare agents. Tests have demonstrated that the solution can decontaminate surfaces contaminated with up to 2.5×10^8 *B. anthracis* spores after a 15-minute contact time. Usually, a disadvantage with wet methods is that the decontamination chemicals are corrosive to materials such as metals, plastics, rubber, paint, leather, and skin. Thus they are not suitable for the use on sensitive equipment. On the other hand, Sandia states that their foam is non-toxic and noncorrosive.

Ionizing radiation (γ-ray) is effective to decontaminate a large area in a heavily shielded fixed facility; for instance, it was applied to decontaminate the post office contaminated in 2001 bioterror attack. However, it was a slow process (took about a year) and very costly. Furthermore, it is not easily implemented in the field and can be destructive to sensitive equipment.

Therefore, alternative dry methods are being developed for many reasons, which include the need to be easily transported, fast working, no mass storage requirement, safe to personnel and inert to sensitive equipment. CAPs carry radical species, which provide bactericidal effect; and the biochemical processes could achieve a high bacteria kill rate, and have no apparent side effects.

5.2 Spores

Bacterial spores are highly resistant to chemical and physical agents. Processes designed to achieve sterilization of food, pharmaceutical, medical, and other products have to thus, of necessity, take this high level of resistance into account. Moreover, commercially available disinfectants are regulated by the Environmental Protection Agency and must be used according to the directions specified on their labels.

A variety of different microorganisms form "spores" or "cysts"; in the vegetative state, each vegetative cell forms one spore and undergoes lysis after the sporulation process is complete. The spore grows exclusively in the mammalian host where spores germinate in the presence of rich conditions such as amino acids, sugars, adequate pH, water, and a favorable temperature.

Some of the spores can change to endospore form in the situation of lack of nutrients. The endospores resist heat, drying, and harsh chemical treatment. In endospore formation, the bacterium divides within its cell wall. One side then engulfs the other. Endospores can lie dormant, which enables bacteria to survive without nutrients for extended periods. For example, dormant spores are able to survive for long periods in contaminated soils and thus account for the ecological cycle of the organism. *B. anthracis*, the etiologic agent of anthrax, is one of those; it is a large, gram-positive, rod-shaped, non-motile, facultative anaerobic, endospore-forming bacterium that causes disease in humans and herbivore animals. The longevity of endospores in the environment is an important factor in the epidemiology of anthrax and explains the predominant occurrence of the disease in herbivores.

B. anthracis bears close genotypical and phenotypical resemblance to *Bacillus cereus* and *Bacillus thuringiensis*. All three species share cellular dimensions and morphology and all form oval spores located centrally in an unswollen sporangium. Because anthrax can cause severe illness and even death, *B. cereus* and *B. thuringiensis* can be used as simulants of *B. anthracis* to test decontamination systems in Labs at low biosafety levels. *B. globigii* was also used in decontamination study. *B. subtilis* is often considered as the Gram-positive equivalent of *E. coli* and has become widely adopted as a laboratory model organism.

An endospore consists of the following main parts: appendages, exosporium, outer coat, inner coat, outer membrane, cortex, germ cell wall, inner membrane, and core. DNA enzymes are contained in

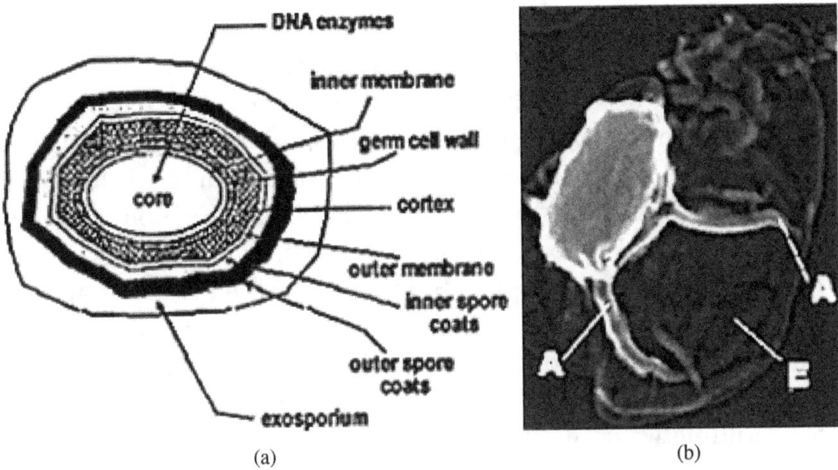

Fig. 5.1. (a) Representation of the interior and exterior structure of a bacterial endospore, and (b) SEM image of a *B. cereus* spore, including its appendages ("A") and exosporium ("E").

the core. The proteins outside the core and the DNA in the core are essential targets of biocides, and the sporicidal effects include protein degradation and DNA damage. Presented in Fig. 5.1a is a representation of the external and internal structure of a *Bacillus* endospore. Figure 5.1b is a scanning electron microscope (SEM) image of a *B. cereus* endospore, in which "A" stands for appendages, and "E" for exosporium.

5.2.1 *Resistance to the treatments*

A spore is highly resistant to a variety of treatments including ultraviolet radiation, pressure, desiccation, high temperature, extreme freezing and chemical disinfectants. When the environment becomes more favorable, the endospore can reactivate itself to the vegetative state. Its coats shield the core from UV radiation; far UV radiation might control the rate of genetic inactivation of isolated microorganisms, but it could not sterilize the aggregated colonies thoroughly and thus UV treatment has little effect on the spore and leaves the spore's immunology almost unchanged. The membranes enable the spore to endure high pressure (100–200 Mpa). Low water content in the core makes the spore heat resistant. The efficiency of the thermal energy treatment method is also limited by the temperature constraint on avoiding damage to equipment and surfaces. Consequently, this method is relatively time-consuming. Moreover, spores are most refractory to inactivation by the boiling water method, which takes about 12 minutes to destroy *B. anthracis* spores. Vigorous boiling could reduce the time to within 3 to 5 minutes to destroy spores from 43 strains of *B. anthracis*. Boiling water in a covered vessel killed spores of the *Bacillus*, reducing the spore population by more than four orders of magnitude in 3 to 5 minutes. Holding water at a rolling boil could further reduce the time to about 1 minute to inactivate waterborne pathogens, including encysted protozoa; however, it would not inactivate the spores even by increasing the boiling time to 3 minutes when an open container was used.

DNA damage is an operative method in killing spores. However, the inner core of the spore consists of various protection mechanisms, including the small acid soluble proteins that are tightly associated with and protect DNA from physical/chemical attack. The spore membranes are also quite rigid structures, where only molecules <200 daltons (Da; 1 Da = 1.66×10^{-27} kg) are capable of passing though. These membranes are important as they contain receptors for various germinant that promote germination and the outgrowth of the spore. On the other hand, if the inner membrane is damaged, dipicolinic acid (DPA) leaks out and germination is subsequently not promoted. Access to these membranes is limited due to the various spore coats at the more external part of the spore structure, which require further damage/penetration in order for these membranes covered inside to be accessed.

5.3 Decontamination via Biological Reactions

One method is to apply a microbial solution to the contaminated area; microbes penetrate the surface to contact and consume the contamination. A detergent or solvent wash is then used to remove the reaction products. The technique could be useful *in situ* to remove hazardous residues in those contaminated areas/contaminators, which include walls and floors, abandoned process equipment, storage tanks, sumps, piping, etc.

Another emerging technology is to use an atmospheric pressure plasma tool for surface decontamination. Air plasma in a highly energized state contains radicals such as excited atoms and molecules. Reactive oxygen species (ROS) and atomic oxygens can destroy just about all kinds of organic contaminants by means of oxidation-reduction reactions, which result to carbonyls and carbonyl adducts and eventually, convert these contaminants into carbon dioxide and water. Oxidation-reduction reactions also lead to the loss of spore viability; these reactions can start at the spore surface to cause protein degradation; the affected proteins and other macromolecules then cross-link each other to consume the spore.

In the following, this emerging technology of using atmospheric pressure plasma as decontaminant is studied. The focus is on the killing of spores. The killing rates of a microwave air plasma for the spore samples in different settings are determined via experimental tests and the killing mechanism is explored and discussed.

5.4 Experimental Preparations

Samples were treated by an arc seeded microwave plasma (ASMP) shown in Fig. 2.4a. The emission spectroscopy of the plasma effluent was examined. As shown in Fig. 5.2, relatively intense spectral lines of OI (in the spectral region between 777.1 and 777.6 nm) indicating high atomic oxygen content in the ASMP were observed. In this measurement, the emission was from a line of sight at 20 mm above the surface of the waveguide (shown in Fig. 2.4a) and the device was run at the airflow rate of 24.6 slm, which was identified via emission spectroscopy to be the optimal condition to produce radicals (mainly atomic oxygen) in the plasma effluent. At this low airflow rate the plasma effluent already has a size with a height of about 25 mm and a volume of about 6 cc and produces an abundance of atomic oxygen.

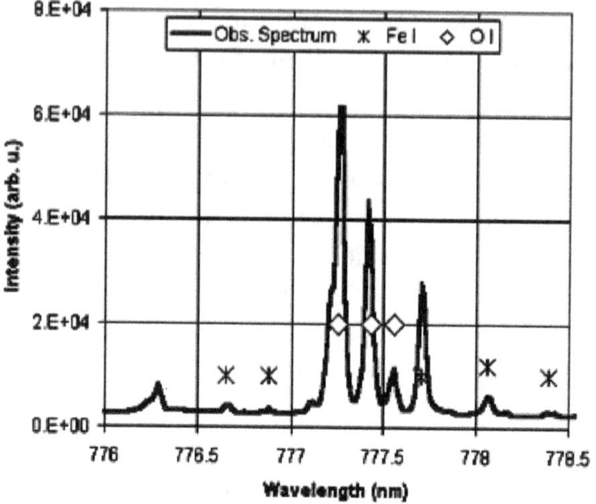

Fig. 5.2. Emission of the plasma at atomic oxygen lines $^5P(J=3)$–$^5S°(J=2)$ at ~777.19 nm, $^5P(J=2)$–$^5S°(J=2)$ at ~777.42 nm, and $^5P(J=1)$–$^5S°(J=2)$ at ~777.54 nm.

5.4.1 *Experimental setup*

ASMP was secured on a fixed vertical lift platform with its 12.5-mm-diameter circular opening on the cavity surface facing downward. A second vertical lift platform directly below the cavity opening was used to position the surface level of the platform below the opening. The plasma effluent was directed downward toward samples placed on the platform. A schematic of the experimental setup is presented in Fig. 5.3.

Three vertical distances of 30, 40, and 50 mm, and exposure times of 2 to 16 seconds were chosen for the dry samples on glass coupons and for the wet samples in 96-well microplates. It was first determined, before the experiments, that at these distances and exposure times, ASMP run at this low airflow rate would not cause noticeable desiccation of water, the sample in a well, using the same arrangement as for the experiments. In the experiments with samples prepared on paper-coupons and placed inside envelopes, a vertical distance of 40 mm was used and the exposure times were limited to be less than 10 seconds. This assured that the plasma

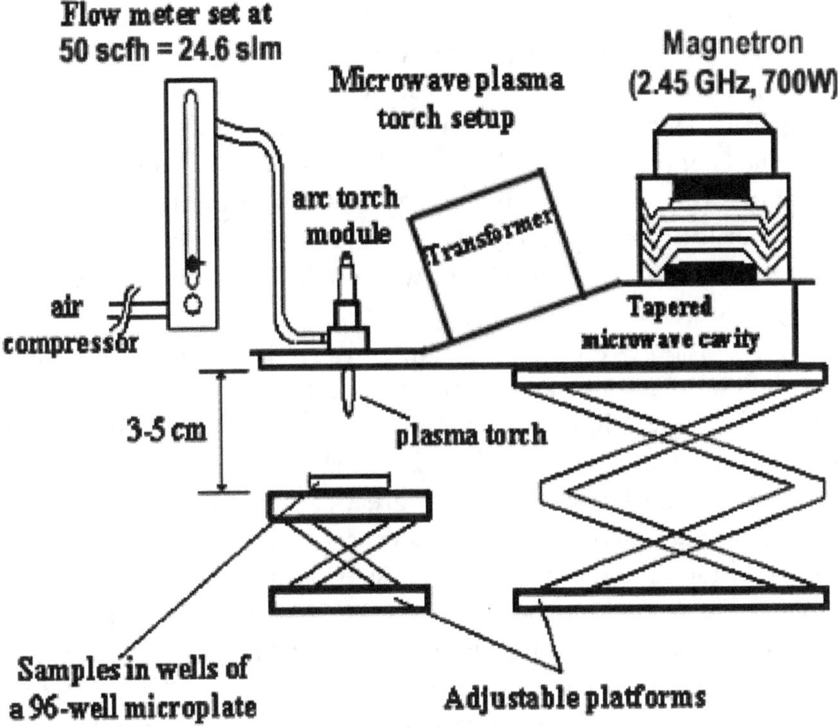

Fig. 5.3. Schematic of the experimental setup.

effluent would not make a noticeable burn mark on the envelope. Exposure times of the samples to the plasma effluent were recorded when ASMP was in full-operation with the magnetron running. The delay of the magnetron operation, typically between 1 to 2 seconds, was ascribed to the necessary heating time of its filament.

5.4.2 *Materials*

B. cereus ATCC 11778 spores were chosen as a simulant substitute for *B. anthracis.* Sterile bacterial spore suspension of *B. cereus* ATCC 11778 (3.5×10^6 spores/0.1 mℓ) and Tryptic Soy Agar (TSA) media were purchased from Raven Biological Laboratories Inc. (Omaha, Nebraska). Water W3500 tissue culture grade was purchased from

Sigma-Aldrich (St. Louis, MO, USA). Gold seal® cover glasses were obtained from Fisher Scientific (Pittsburgh, PA). Organ Tissue Culture dishes were purchased from Falcon Plastics (Oxnard, CA). Anhydrous calcium sulfate® was purchased from W.A. Hammond Drierite Company LTD. (Xenia, OH, USA). Sterile tubes (0.5mℓ) and ultra-dishes Petri were obtained from PGF Scientific (Gaithersburg, MD). Sterile tips were purchased from VWR International (Bridgeport, NJ). Regular legal-size envelopes without sterilization were used to contain samples.

B. cereus spores are aerobic (oxygen is needed for growth), widely distributed in nature, and cause various food spoilage. But these agents are relatively benign microorganisms having the minimum safety requirement (i.e., Biosafety Level 1 (BSL1)) and do not ordinarily cause human disease. However, great care is still required in the preparation and handling of *B. cereus* spores, during and after experiments. These bacterial spores can cause two distinct types of illnesses: 1) diarrhea with an incubation time of approximately 4 to 16 hours, and 2) an emetic (vomiting) illness with an incubation time of 1 to 5 hours. Spore handling was carried out under a Biological Safety Cabinet Class IIA/B3 from Forma Scientific Inc. (Marjetta, OH).

5.4.3 *Sample preparations prior to and after exposure*

5.4.3.1 *Prior to exposure*

A. Dry samples on glass slide-coupons

Glass coupons were prepared by fixing a glass slide to the Organ Tissue Culture dish and exposing it to UV irradiation for 5 hours to prevent cross contamination. Dry samples were prepared by inoculating spore suspension of 30$\mu\ell$ (i.e., ~10^6 spores) onto each of the UV irradiated glass slide-coupons; it forms a spot of ~5 mm in diameter on the glass slide. These samples were then desiccated for 12 hours.

B. Wet samples

Drops of bacterial-spore solution were inoculated onto the wells of 96-well microplates. Each drop of 30$\mu\ell$ was taken directly as a wet sample.

C. Dry Samples on paper-coupons

A fiber mixed (DuPont™ Tyvek) envelope was cut into 10×10 mm squares to host spore samples. Each paper-coupon was fixed to the Organ Tissue Culture dish and exposed to UV irradiation for 5 hours to prevent cross contamination. Samples were prepared by inoculating spore suspension of 30 $\mu\ell$ onto each of the UV irradiated paper-coupons, which were then desiccated for 12 hours. Subsequently, each coupon was inserted into an envelope. Envelopes were sealed before being exposed to the ASMP and one of those was kept as a control. Spores were in a *"confined environment"*.

In these sample preparations, every sample contained about 10^6 spores before the treatment.

5.4.3.2 *Procedures of processing the samples after plasma treatment*

A. For dry samples on the glass coupons:

Treated spores and debris were removed from the glass slides by means of extensive sonication using tissue culture water. The mixtures were serially diluted from 10^{-1} to 10^{-7}.

B. For wet samples in the 96-well microplates:

Volumes of solutions of untreated and post-exposed spore samples were checked to make sure that the 30 $\mu\ell$ dripped initially into each well of the plate was not reduced by more than 5% (controlled by the exposure distance and time). The post-exposure sample was handled as follows: spores and debris in each well of the plate were diluted using W3500 tissue culture water (60 $m\ell$/well) and mixed by means of extensive continuous shaking for 1 hr at 25°C. The mixtures of post-exposed samples were serially diluted from 10^{-1} to 10^{-5}.

C. For samples on the paper-coupons:

Treated paper-coupons hosting spores and debris were placed into organ tissue culture dishes. Then, 1 $m\ell$ of tissue culture water was added to each dish. Soon after, each dish with coupon was extensively sonicated for 20 minutes at 20°C to remove unbound spores and debris from the paper-coupon. The mixtures were serially diluted from 10^{-1} to 10^{-7}.

In all cases, those diluted mixtures were plated onto petri dishes with Tryptic Soy (TS) liquid media and incubated at 37°C for 16 hours. Untreated control was sonicated and handled in the same manner as treated samples. After the incubation, the resulting CFU could be observed through their images in the pictures that were taken.

5.5 Colony-forming Unit (CFU)

A colony-forming unit (CFU) is a unit used to estimate the number of viable bacteria or fungal cells in a sample. A cultured viable cell grows in replication to form a visible colony unit. The method of counting CFU to estimate the number of viable cells assumes that every colony is separate and founded by a single viable microbial cell. Therefore, it is necessary to dilute the sample to the linear range (less than 300 CFU on a standard sized Petri dish) before incubating CFU, ensuring a colony is replicated by a single viable cell.

The spread plate method spreads the sample (in a small volume) across the surface of a nutrient agar plate (petri dish), which is allowed to dry before incubation for counting.

Typically seven-fold dilutions are used; the concentration of the sample is serially diluted until sample plates with countable number of CFUs can be prepared. It is also realized that if the number of CFUs in a sample plate is too low after a serial dilution, it can give large statistical error. Therefore, the dilution only in the chosen range (based on the understanding; it may go through trial and error iterative procedure) is plated on a petri dish with 2 or 3 replicates, for incubation. The visual appearance of a colony in a cell culture requires significant growth; however, the incubation time has to be controlled to ensure each individual CFU is identifiable. A picture of each plate is taken by a digital camera and colonies can be enumerated from pictures of plates using software tools. The CFU/plate read from a plate is then compared with the original viable cells/plate deduced mathematically from the amount in the control sample, factoring in the dilution factor, to evaluate the drop, logarithmically.

The samples in three different settings encounter plasma effluents differently, it is expected that the killing rate decreases from case (A) to case (C). Thus, the mixtures with 10^{-1}, 10^{-2}, and 10^{-3} dilution in cases (A), (B), and (C), respectively, were chosen in counting CFU. The results of the counted CFU in each case were then compared with the number of corresponding control CFU (about 10^5, 10^4, and 10^3 per diluted sample before the treatment in cases (A), (B), and (C), respectively) to determine the decontamination efficacy (i.e., the survival curves, called "kill curves" in some of the literature). To ensure full spore removal from the paper-coupons, after sonication, those paper-coupons were placed onto Petri dishes with TSA media for incubation. No CFU was observed on sonicated paper coupons.

5.6 Decontamination Experiments and Results

5.6.1 *Dry samples on glass slide-coupons*

The efficacy of the ASMP on decontamination of *B. cereus* is demonstrated in Fig. 5.4, which contains images representing the results of 10^{-1} diluted treated samples. As shown, the number of CFUs in each petri dish drops rapidly with the increase of the plasma treatment time. Almost all spores were killed in about 6 seconds at 30 mm exposure distance, and in 12 seconds at either 40 or 50 mm exposure distances. The CFU counts N (viable spores remaining) from the experimental results (each set of experiments was run twice) were normalized to the initial number N_0 (CFU control number). These data points are presented in Fig. 5.5 and are fitted by straight lines as the kill curves for dried *B. cereus* spores exposed to the plasma effluent at three exposure distances: 30 mm (+), 40 mm (Δ), and 50 mm (*).

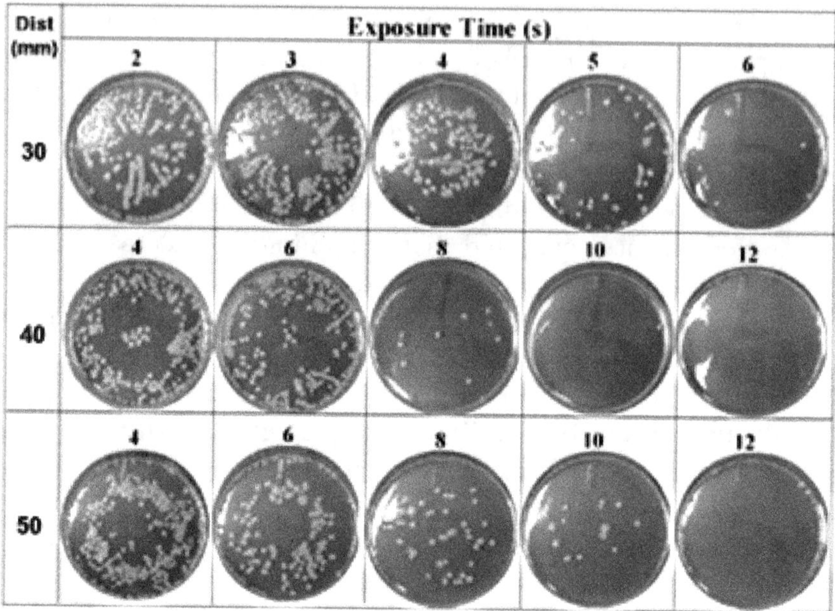

Fig. 5.4. CFU formation after decontamination using ASMP at several distances (30, 40, 50 mm) and exposure times (2–12 seconds).

In Fig. 5.5, the horizontal-axis is the exposure time in seconds and the vertical-axis is the normalized viable spores remaining in decimal log, i.e., $\log_{10}$ (N/N_0). The percentage of total spores remaining is plotted in the range down to –5 decimal logs. The results are compared with previously reported decontamination of *Bacillus globigii* (BG) spores using APPJ (dashed line) and hot gas of 175°C (dotted lines).

The APPJ kill curve was obtained for the samples' exposure distance set at 5 mm. The time required to reduce the viable (BC) spore population by a factor of 10 by the ASMP in this graph for 30, 40 and 50 mm distances are calculated to be 2.09, 2.84, 4.08 seconds, respectively. For comparison, the APPJ and hot gas exposure times required to reduce the viable (BG) spore population by the same factor are calculated to be about 4.5 and 45 seconds. Taken together, the experimental results have shown that all spores were essentially killed in less than 8 seconds at 30 mm distance, in 12 seconds at 40 mm, and

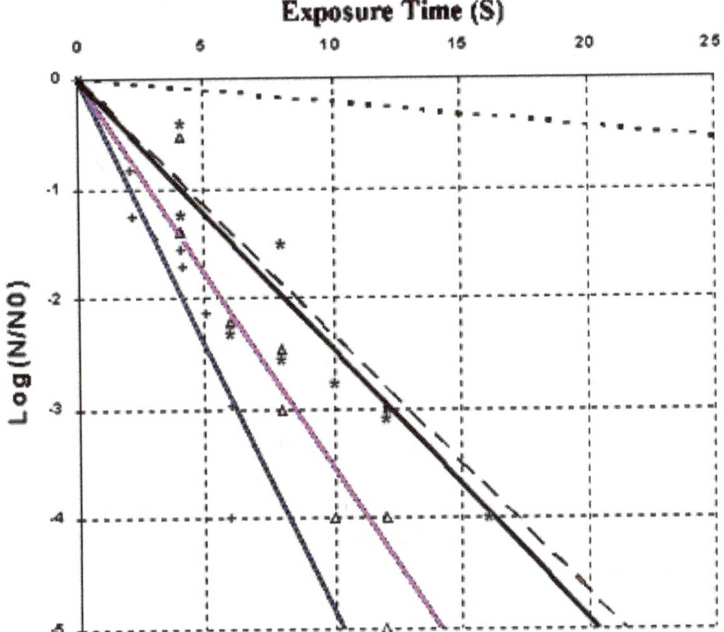

Fig. 5.5. BC kill curves: data points are obtained by placing samples at three exposure distances of 30 mm (+), 40 mm (Δ), and 50 mm (*) from the cavity opening of the ASMP. Results are compared with BG kill curves reported by Herrmann *et al.* (Phys. Plasmas **6**, 2284, 1999), using APPJ (dash line) and hot gas of 175°C (dot line).

in 16 seconds at 50 mm distance away from the opening of the microwave cavity of the ASMP. In comparison, it took about 25 s for APPJ to achieve a thorough kill of BG spores.

5.6.2 *Wet samples*

B. cereus spores in solution were exposed to the plasma effluent at three exposure distances of 30 mm (●), 40 mm (○), and 50 mm (+). The sterilizing time was increased from 2 to 16 s. The percentage of total spores remaining is plotted in the range down to −1 decimal log. The fitting lines of the data points, representing the kill curves, are presented in Fig. 5.6. The time required to reduce the viable B. cereus spore population by a factor of 10 by the ASMP in this graph for 30, 40 and 50 mm exposure distances

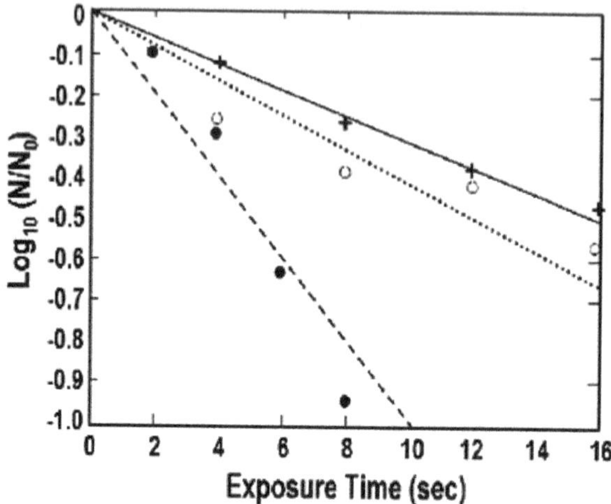

Fig. 5.6. BC kill curves; data points are obtained by placing wet samples (*B. cereus* spores in water) at three exposure distances of 30 mm (•), 40 mm (○), and 50 mm (+) from the cavity opening of the ASMP.

are calculated from the fitting lines in Fig. 5.6 to be 10, 24, and 31 seconds, respectively.

5.6.3 *Sample contained inside an envelope*

The ability of the ASMP to kill bacterial spores contained inside an envelope was tested. It was found that the plasma effluent of the ASMP killed about 85% of spores in 7 seconds.

The CFU counts N from the experimental results were normalized to the initial number $N_0 \cong 1000$ (10^{-3} diluted control sample). These data points (averaged over two sets of data) are presented in Fig. 5.7 and are fitted by a straight line as a kill curve. The percentage of total spores remaining is plotted in the range down to −1 decimal log. The time required to reduce the viable (BC) spore population by a factor of 10 by the ASMP in this graph is calculated to be about 9 seconds.

5.7 Morphological Studies

In the wet-sample experiments, many viable spores still remained after the treatment. Thus the treated wet samples (of 10^{-2} dilution)

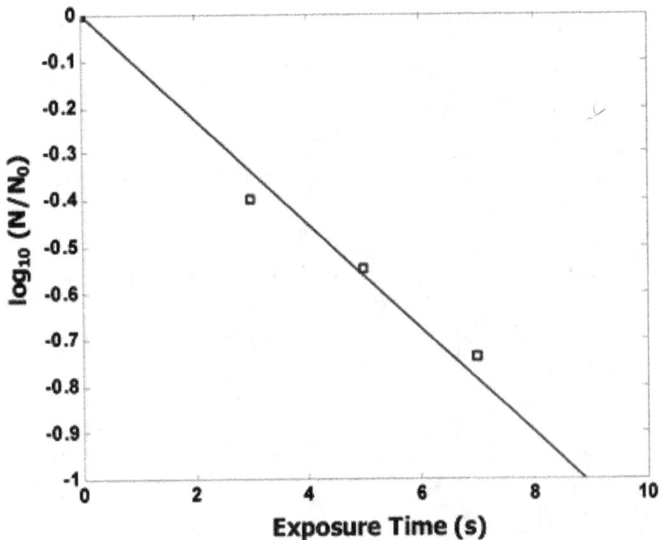

Fig. 5.7. BC kill curve: data points are obtained by placing each envelope containing a sample at an exposure distance of 40 mm from the cavity opening of the ASMP.

at 30 mm distance were used in the scanning electron microscope (SEM) and atomic force microscope (AFM) studies discussed in the following. SEM produces a two-dimensional (2-D) image to reveal the actual shape and the morphological structure of a bacterial spore, while AFM examines the spore's cell properties, including shape, structure, dimensions, and volume, by taking a three-dimensional (3-D) image.

5.7.1 *Scanning electron microscopy*

Solutions of untreated (10^4 CFU) and post-exposed spore samples for SEM observations were deposited on mica disks and desiccated for 7 days. Samples were then coated with a 10 nm thin film of evaporated gold for 60 sec and then observed with a SEM at an accelerating voltage of 15–20 kV. The images of untreated (Fig. 5.8a) and exposed but still viable *B. cereus* spores (Figs. 5.8b–d) were taken at low (left column) and high (right column) magnification for examination.

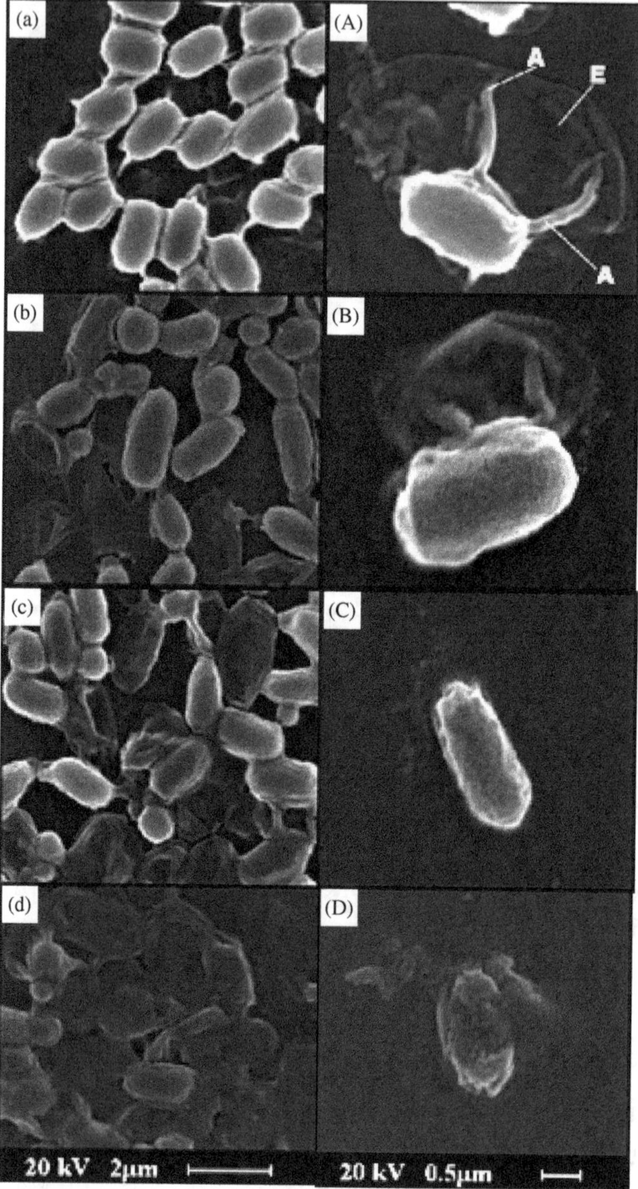

Fig. 5.8. SEM images of *B. cereus* spores; (a) untreated and (b)–(d), exposed to the ASMP at 30 mm distance for (b) 2 s, (c) 4 s, and (d) 8 s at low (left column) and high (right column) magnification. In (a), A stands for appendages, E for exosporium.

Exosporium (labeled by E) and appendages (labeled by A) of a spore are clearly seen in Fig. 5.8a. Comparing images (Figs. 5.8b–d) of individual spores (right column) exposed to the plasma effluent for different time periods (2 to 8 sec) with that of the control reveals the changes of the actual shapes and morphological structures of bacterial spores during exposure to the ASMP. The images of grouping spores in the left column show the integrity of spores. In the case of the 2 sec plasma treatment shown in Fig. 5.8b, spores remained intact to keep their integrity for the most part; the exosporium and appendages of a spore are still visible, but its size seems to increase considerably from that of the untreated spore (Fig. 5.8a). Spores after 4 sec plasma treatment show changes in morphology. As seen in Fig. 5.8c, the spore has lost most of its appendages, its exosporium has shrunk and its size has decreased drastically. However, the actual change of a spore's size has to be further checked by the corresponding 3-D images, which will be presented in the following section. Increase of the plasma treatment time to 8 sec has drastically changed the morphology of the spore. Fig. 5.8d indicates that spores have lost integrity; neither appendages nor exosporium of a spore can be seen.

5.7.2 *Atomic force microscopy*

Solutions of untreated sample (10^4 CFU) and those after exposure (for 2, 4, and 8 s) were immobilized on mica discs using sterile syringes, and then dried in air at 20°C. All spores were found to be firmly attached to the mica disk and remained sufficiently bound to be imaged. Prepared samples were later mounted on an AFM sample holder for imaging. All AFM observations were carried out at 20°C, using a Nano Scope® IIIa controller as well as a MultiMode™ microscope operating in tapping mode (amplitude) together with an E-scanner. A 125-µm silicon nano-probe was also employed. The calculated spring constant was 0.3 N/m. The resonance frequency remained in the range of 240–280 kHz, and the scan rate was 1 µm/s.

Flattening and high-pass filtering of the image data were performed to remove the substrate slope from images and high-frequency noise strikes, which are, otherwise, more pronounced in the high-resolution tapping mode imaging.

Images of untreated *B. cereus* spores and three treated spore samples corresponding to the three exposure times, 2, 4, and 8 s, are presented in Figs. 5.9a–d, in both low resolution 10.0 × 10.0 µm (column A) as well as in high resolution 3.0 × 3.0 µm (column B). The images of untreated spores in Fig. 5.9a indicate that the outer membrane of *B. cereus* spore has an elongated (rod) shape. The high-resolution image in column B exhibits clearly identifiable appendages (labeled by A) and are covered with a loose layer of exosporium (labeled by E) that has slightly spread and is stuck to the mica surface. As seen in column B of Fig. 5.9, the cell of the untreated one (a) has a bubbling shape in the middle region. After the exposure, the cell is squashed in the middle region and becomes elongated and wider as seen in (b) to (d).

The middle part of the cell is flattened and the flattened region expands toward the two ends as the exposure time increases from 2 s to 8 s. That there is no apparent damage of the outer spore coat can be seen; it suggests that protein degradation may be a likely inactivation mechanism; it causes the collapse of the external structure of the spore. Again, the actual change of the size of the spore, in particular, of the volume of the spore, has to be further checked by the corresponding 3-D images.

The corresponding 3-D images of the spores, presented in column B of Fig. 5.9, are presented in the first row of Fig. 5.10. It is noted that these images have different spatial scales. The axes on the horizontal plane in (b) and (c) extend to 5 µm from 3 µm in the other two figures in (a) and (d). The vertical axis in (a) extends to 2 µm from 1 µm in the other three figures in (b), (c), and (d). AFM allows section analysis in virtual 3D. This provides accurate length (in second row labeled by A), width (in third row labeled by B), and height (in fourth row labeled by C) of an observed specimen.

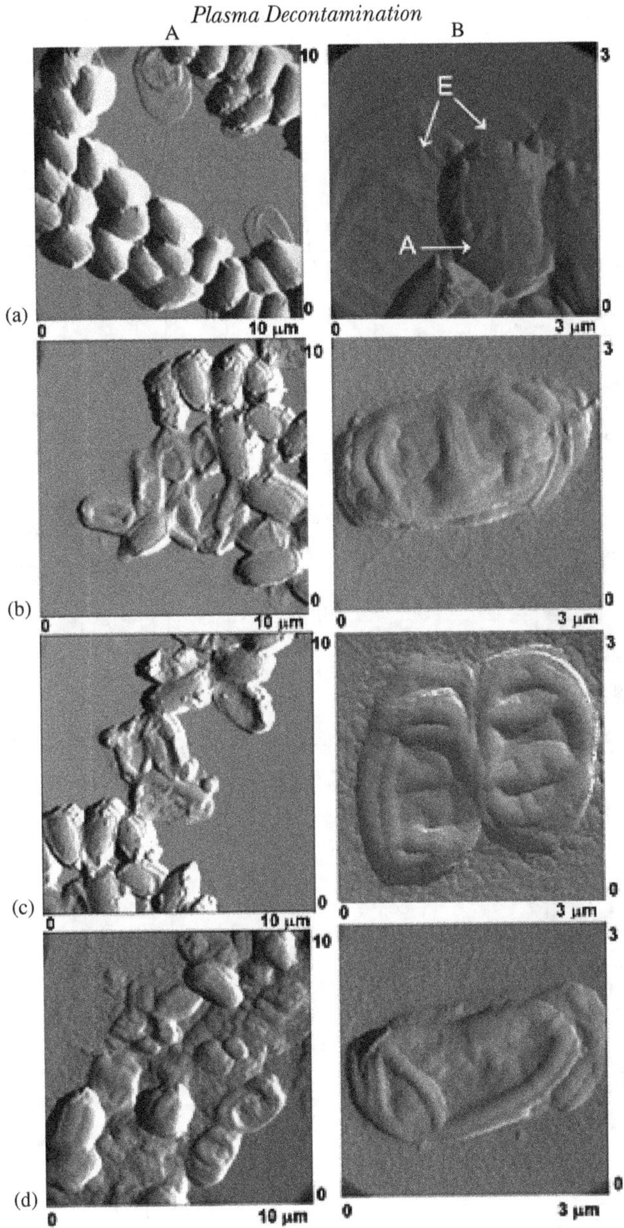

Fig. 5.9. AFM amplitude images of *B. cereus* spores at low (column A) and high (columns B) resolution; (a) untreated and (b)–(d), exposed to the ASMP at 30 mm distance for (b) 2 s, (c) 4 s, and (d) 8 s.

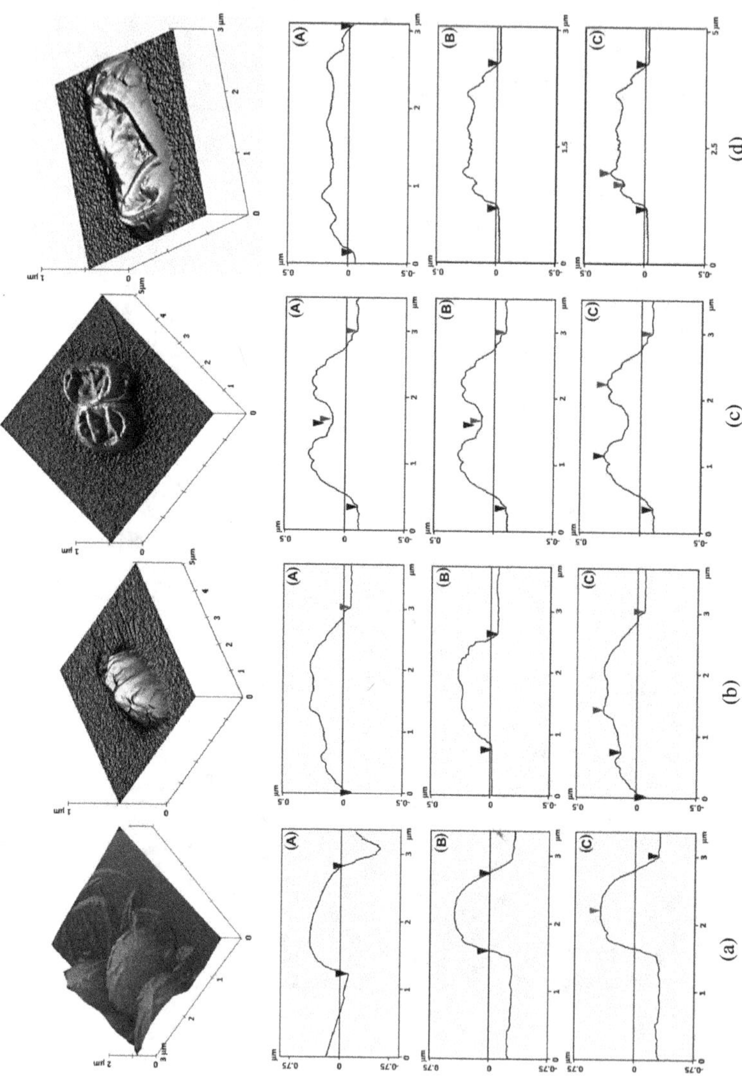

Fig. 5.10. AFM 3D images (first row) and section analyses for the length (A), width (B), and height (C) of (a) untreated *B. cereus* spore and of the viable spores after plasma treatment for (b) 2 s, (c) 4 s, and (d) 8 s.

Section analysis (SA) reveals the approximate dimensions, "length" × "width" × "height", of the spores shown in the first row to be:

(a)   Untreated spore: 1.6 μm × 1.1 μm × 0.734 μm.
(b)   After 2 seconds treatment: 2.9 μm × 1.8 μm × 0.35 μm.
(c)   After 4 seconds treatment: 2.65 μm × 1.3 μm × 0.375 μm.
(d)   After 8 seconds treatment: 2.8 μm × 1.8 μm × 0.275 μm.

Spore dimensions have changed significantly: a 175% increase in length (from 1.6 μm to 2.8 μm), a 164% increase in width (from 1.1 μm to 1.8 μm), but a 267% decrease in height (from 0.734 μm to 0.275 μm). This morphology change can be best described as a structural depletion or collapse via oxidation-reduction reactions (i.e., rapid growth as well as protein degradation stimulated by the atomic oxygen without support of nourishment).

Specifically, hydrogen peroxide (H_2O_2), generated via the reaction of atomic oxygen with H_2O, is a cell signaling molecule, which can react with small, acid-soluble proteins in the coats. This marked effect on the structure of spore coats results in the depletion or absence of the cortex, and in the disorder of the ribosomes, explaining the collapse of spore coats. However, activation of peroxide to hydroxyl radicals (*OH) is necessary for sporicidal action. The hydroxyl radical is the strongest oxidant known, and it is by this activation that hydrogen peroxide can do the actual killing of bacteria. The hydroxyl radical, being highly reactive but having a very short in vivo half-life of approximate 10^{-9} s, can attack membrane lipids, DNA, and other essential cell components.

5.8 Plausible Mechanism

A thermocouple probe (Omega model: DP460) was directly exposed to the ASMP (at a distance $\geq$30 mm), the temperature elevation did not exceed 45°C. This thermocouple was also used to check the temperature elevation of a water drop after exposing it to the ASMP. The probe was cover by the water drop in a petri dish, where the volume of the water drop was larger than that of the solution in each well of the microplate; nevertheless, the temperature increase was negligibly small. Less than 5% of the sample solution was vaporized after exposure to the ASMP, further suggesting that the temperature of the ASMP at the sample location could not be high. Furthermore, the probe was inserted inside an envelope which was placed at the same distance (40 mm) as that of the treatment experiments below the cavity opening of the ASMP. With 8 seconds of exposure to the ASMP, it was found that the temperature was only raised to 38°C from the room temperature of about 26°C. The measurements showed that the heat applied to the samples in all experiments was too low to directly kill spores. It is concluded that thermal process as the decontamination mechanism can be ruled out in all experiments.

The diameter of the circular opening of ASMP on the microwave cavity surface is about 12.5 mm, which is much smaller than the wavelength of about 123 mm; hence, the evanescent fields leaking out of the cavity were too weak to have any impact on the spores. For the safety reason, a microwave leakage detector (MD-2000) was used in experiments to monitor the level of the microwave flux. It was found that the power flux at 1 m distance away was less than 10 $\mu W/mm^2$, which was within the safety threshold level of 50 $\mu W/mm^2$. Thus we also rule out the possibility that leaked microwave radiation or the evanescent fields could be responsible for killing spores as well as for the changes on the morphological characteristics of the spore.

Chemically reactive oxygen species (ROS), such as atomic oxygen, molecular singlet oxygen and ozone are known to be effective in the destruction and annihilation of bacterial spores, and atomic oxygen is probably the most effective one among them in decontamination. Four likely processes to produce atomic oxygen are presented in Sec. 3.1. Among those, the process of dissociative attachment of electron to molecular oxygen, $e^- + O_2 \rightarrow O + O^-$, and the process of dissociative recombination of electron with molecular oxygen ion, $O_2^+ + e^- \rightarrow O + O$, are likely the main processes to generate atomic oxygen in low temperature air plasma.

Seed plasma produced by arc discharge can quickly absorb microwaves to generate non-equilibrium microwave plasma, which becomes the catalyst to produce atomic oxygen in the airflow. This was verified by the emission spectroscopy of the ASMP. It was found that there were two dominant groups of lines in the emission spectra. One was from metallic contaminants, predominantly Fe and Cu, the material of one of the electrodes (frame of the arc discharge module) and the copper cavity, present in the form of particulates in the solid phase. The other one, in the spectral region between 777.1 and 777.6 nm as shown in Fig. 5.2, was the atomic oxygen (OI) lines. Molecular emission spectrum of Nitrogen and Oxygen was practically nonexistent, compared with the spectral intensity of the OI lines and metallic lines. The emission spectrum in the UV range was also checked. uv radiation was detected, but its intensity was not strong, compared with that of the metallic lines. The *OH line at 305 nm was buried in the metallic lines.

The intense OI spectral lines shown in Fig. 5.2 indicate relatively high atomic oxygen content in the plasma effluent. Moreover, the large kill rate on the dry samples suggests that the ASMP carries an abundance of atomic oxygen.

However, the lifetime of the atomic oxygen is about 0.2 to 0.3 ms. With a flow speed of 20 m/s, most of produced atomic oxygen in the discharge can only survive a travel distance of about 4 to 6 mm. Therefore, the 30 to 50 mm exposure distance is reachable only by locally produced atomic oxygen.

As seen in Fig. 5.11, the arc loops, (A) and (B) without, and (C) with, the presence of microwaves, of the discharges (as the source

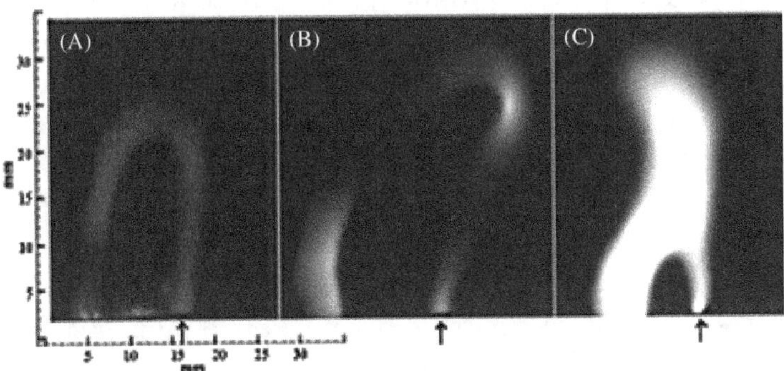

Fig. 5.11. Images of arc loops in ASMP taken by an intensified CCD camera set for a 100 ms exposure time; arrows indicate the position of the central electrode.

region of the atomic oxygen) extend out from the cavity surface by nearly 30 mm. The microwave has significantly intensified the discharge. This explains why ASMP can still effectively decontaminate at an exposure distance as large as 50 mm.

Atomic oxygen can react with water to produce hydrogen peroxide (H_2O_2), which is then activated to hydroxyl radicals (*OH) for sporicidal action. ASMP contains metallic particulates; thus, another possibility was *OH radicals formed in water through collision of metallic particulates with water molecules.

Oxidization of *B. cereus* bacterial spores by reactive radicals generated in the plasma effluent (such as OI) or in water (such as H_2O_2 and *OH) is then suggested to be the mechanism of killing spores in the experiments. This non-thermal destruction mechanism primarily involves the chemical reactions of ROS with nucleic acids, lipids, proteins and sugars. Active oxygen species and singlet oxygen cause extensive oxidative damage to biological macromolecules including lipids, carbohydrates, and proteins. These chemical reactions result in the production of carbonyls and carbonyl adducts. Consequently, the morphology and physical shape of a spore are changed.

The 2-D and 3-D shapes of the *B. cereus* bacterial spore before exposure (Fig. 5.8a and Fig. 5.10a) and after exposure to the plasma effluent (Figs. 5.8b–d and Figs. 5.10b–d) are recorded by SEM and AFM, respectively, for comparison. It appears that the shape of the exposed spore in Fig. 5.8c becomes longer and narrower than that of the untreated spore shown in Fig. 5.8a; moreover, destruction of the spore's morphology is demonstrated in Fig. 5.8d. The cells of the similarly exposed spores, shown in Figs. 5.9c and d, are squashed and flattened. Such changes on the spore's morphological structures and shape add support to the suggestion that it is the oxidation agent (atomic oxygen) in the plasma effluent that destroys the spores via biochemical processes.

The kill time (i.e., 10-fold reduction time) for the wet sample is about 10 seconds at an exposure distance of 30 mm, 24 seconds at 40 mm, and 31 seconds at 50 mm (Fig. 5.6). These times are much longer than the corresponding ones (Fig. 5.5) of 2, 3, and 4 seconds for the dry samples, which are directly in contact with the plasma effluent. This is understandable because the atomic oxygen in the plasma effluent had to pass through a thin water barrier before directly reacting with the spores; and probably, the densities of H_2O_2 and *OH radicals formed in water were relatively low.

The kill time for the paper-coupon sample placed underneath the double-layered glue area of an envelope is about 9 seconds (Fig. 5.7). It reduces to a half if only a single paper layer covers the sample.

Based on direct microscopic observations of plasma exposed bacterial endospores, with exposure to a greater amount of air plasma, more collapse of spores is observed. There is a significant relationship between the degree of collapse and spore death through the dependence on the exposure time. The sporicidal mechanism of air plasma is likely through the oxidation reduction process activated by the atomic oxygen carried in the air plasma. By contrast, N_2 plasma as well as noble gas plasmas do not cause any collapse/shrinkage of endospores. The sporicidal mechanisms of air plasmas and inert (N_2)/noble (He, Ar, etc.) gas plasmas are clearly different, but air plasmas appear to have more extensive effects on bacterial spores and much higher kill rate.

Currently, various methods are used in decontamination. These include physical methods, such as heat and radiation sources, as well as a range of chemicals, including hydrogen peroxide, glutaraldehyde and alcohols. The presented tests, on the other hand, have demonstrated that air plasmas can be an effectual dry decontaminant.

5.9 Applications in Other Emerging Areas

Plasma decontamination can also be applied for wastewater treatment and in the food industry.

5.9.1 *Water treatment*

As wastewater remediation becomes a global concern, the development of innovative advanced oxidation processes for wastewater treatment is a major challenge. Electric discharge plasma has been applied for decomposing a range of pollutants and introducing oxidation.

The purity of the water, such as dissolved gases which form micro-bubbles in the water, plays a significant role in the breakdown process. The breakdown voltage is given by

$$V \geq \left(\frac{DC_p \rho T_0^2}{\sigma_0 E_a} \right)^{1/2} \left(\frac{L}{R_0} \right) \tag{5.1}$$

where V, D, $C_p \rho$, T_0, σ_0, E_a, L, and R_0 are the breakdown voltage, thermal diffusivity of water (ca. 1.5×10^{-7} m^2/s), specific heat per unit volume, temperature, water conductivity, Arrhenius activation energy for the water conductivity, breakdown channel length, and breakdown channel radius, respectively.

A typical order of magnitude of the local electrical breakdown field of water is 1 MV/cm (in the case of microsecond pulsed breakdown), which is more than 30 times the breakdown electrical field (~30 kV/cm) of atmospheric pressure air.

The ROS emission lines at 309 nm (OH) and at 777.4 and 844.6 nm (OI) are detected. Thus OH radicals and H_2O_2 generated through chemical processes provide biocidal effect, which destructs bacterial membrane, disrupts enzymatic activity of bacteria, and affects both purines and pyrimidines in nucleic acids.

When plasma technologies are combined with catalysts, synergetic pollution abatement results may be achieved. Further research includes the study of the efficacy of different electric discharge types and methods, as well as system designs.

5.9.2 *Food industry*

In the food industry, intensive production can cause mishandling of food processing, transport and storage, which may increase the level of mycotoxin contamination, resulting in human diseases affecting vital systems such as the nervous and immune systems. From the CDC, each year roughly 1 in 6 Americans (48 million people) get sick, 128,000 are hospitalized, and 3,000 die of food-borne diseases. Efficacy of the available decontamination methods to appropriately deal with food-borne diseases is of interest for food manufacturers as well as for the safety of consumers. Of course, it is also preferable that the methods have the capacity of being a dry rapid process, producing no toxic residuals and adverse effects, consuming low energy, decontaminating solid foods, preventing biofilm formation, and advancing pasteurization methods.

CAP offers a novel non-thermal and dry decontamination technology, which can be used to inactivate contaminating microbes on foods, as well as to sanitize processing equipment and to disinfect the environment of food processing areas. It works to reduce the need for the use of chemicals and water, which make this technology cost effective and ecologically neutral. Moreover, it has less effect on the quality of food products following treatment compared to classical methods.

Plasma treatment deposits barrier coatings onto homopolymeric packaging materials to replace multi-layer polymer packaging materials. It reduces the materials used, which in turn, reduces the cost and weight, and improves the recyclability. Moreover, plasma treatment removes unwanted 'organic contaminants' and increases wettability and adhesion, which provides anti-microbial action and improves printability and anti-mist formation of packaging materials.

Further research is needed to scale-up plasma technology to make it available for industrial processes. Optimization and scale up to commercial sanitizing levels require a more complete understanding of involved antimicrobial chemical processes and the effects on the sensory and nutritional qualities of treated foods. The variety and complexity of the necessary equipment to meet industrial requirements are challenges of the development.

After all, it is necessary to meet food safety regulations and industry standards in the decontamination of food and food contact surfaces. Test results suggest that CAP treatment can meet the requirement, but its generator also has to be easy to implement into existing processing lines, compatibility with existing sanitation practices and rapid decontamination in order to be cost-effective.

Another potential area of application is related to the control of the properties of food. Proteins, as a food constituent, play a remarkable role in the techno-functional characteristics of processed foods and/or the physicochemical properties of protein-based films. At the same time, some proteins are responsible for reduction in quality and nutritional value, and/or causing allergic reactions in the human body. Therefore, the influences of different types of CAP on the conformation and function of proteins with food origin, especially enzymes and allergens, as well as protein-made packaging films, will be of interest.

Different active species of CAP can cause different effects on the enzyme structure and its function. The level and type of variations in the functional properties of food proteins, purified proteins in food, and CAP-treated protein films are also affected by a number of control factors, including treatment power, time, and gas type of CAP, as well as the nature of the substance and the treatment environment. Considerable effort to explore these parameters is called for.

In sum, the specific research areas include microbial decontamination of food products and packaging material processing, functionality modification of food materials, and decadence of agrochemical residues; additional areas include hydrogenation of edible oils, mitigation of food allergy, inactivation of anti-nutritional factors, tailoring of seed germination performance and effluent management. It is aimed at establishing plasma processing as an eco-friendly technique with minimal changes to food products, making it a befitting alternative to traditional techniques. It is called for active researches to up-scale CAP technology to a commercial application level, and to establish operation protocols to meet food safety regulations and industry standards in the decontamination of food and food contact surfaces.

Problems

P5.1. A variety of chemical, biological, and radiological weapons are collectively known as HAZMAT/weapons of mass destruction (WMD). Incidents involving HAZMAT/WMD are complicated because victims may become contaminated with the hazardous material. Decontamination technologies are being developed to neutralize, destroy, or otherwise mitigate a chemical hazard to meet the aim of decontamination which is to make an individual and/or their equipment safe by physically removing toxic substances quickly and easily. CAAPs are dry decontaminants. List the likely advantages of dry decontamination over liquid and foam decontamination.

P5.2. Simulants are organisms or substances which mimic physical or biological properties of real biological agents, without being pathogenic. They are used to study the efficiency of various dissemination techniques or the risks caused by the use of biological agents in bioterrorism. For instance, *Bacillus subtilis* and *B. cereus* which causes food poisoning are valuable models for research on anthrax decontamination.

Use *Bacillus subtilis* as a model and apply fluorescent microscopy techniques to show the differentiation, gene/protein regulation, and cell cycle events in bacteria.

P5.3. Explain the difference between sterilization and bio-decontamination.

P5.4. The boiling point of H_2O_2 is 150°C. Vaporized hydrogen peroxide (VHP) is one of the dry decontaminants. The use of VHP has some great advantages compared to liquid spray on. It can be distributed faster in a large volume and penetrates better than sprayed liquid. It also requires less aeration time, because less liquid has to be introduced into the volume. Moreover, it leaves no residue on surfaces because it decomposes to water vapor and oxygen. Thus VHP does not have to be wiped down after bio-decontamination. It is widely used to kill microorganisms.

Measure D values of VHP on *Bacillus cereus* in time to plot a VHP kill matrix. Place the stainless steel coupon inoculated with *Bacillus cereus* spore sample in a glass ware containing concentrated (35% or 59%) hydrogen peroxide liquid (Vaprox). The partially covered glass ware is heated at 30°C to vaporize Vaprox.

P5.5.　Compare the sporicidal efficacy of atomic oxygen (OI) and VHP. [Hint: the sporicidal efficacy of OI can be extracted from the data presented in Fig. 5.5]

Chapter 6

Effect of Air Plasma on Blood Coagulation

Blood is a fluid tissue that includes 60% of a liquid portion known as blood plasma, and 40% of formed elements or blood cells. Blood plasma, a protein-salt solution up to 95% of water by volume, suspends red blood cells (RBC), white blood cells (WBC), and platelets alike. It contains albumin (the chief protein constituent), fibrinogen (responsible in part for blood clotting), globulins (including antibodies) and other clotting proteins. Formed elements consist 86.6% of RBC, 10.4% of platelets, and 3% of WBC. RBC are produced in bone marrow and contain hemoglobin, a complex iron-containing protein that carries oxygen and participates in carbon dioxide exchange. WBC consist of neutrophils, eosinophils, basophils, monocytes, and lymphocytes; they are involved in destroying microorganisms and produces antibodies that help the body fight infection. Platelets play a vital role in the early response to vascular injury; at a wound site when in contact with air, platelets become activated, agglomerate, and form platelet clots adhering to injured blood vessel wall components; platelets also secrete mediators that attract WBC.

The average lifetime for WBC is hours to days, for RBC 120 days, and for platelets 9 days. Platelets are more fragile cells and 3–4 times smaller than RBC. The composition of whole blood is depicted in Fig. 6.1a.

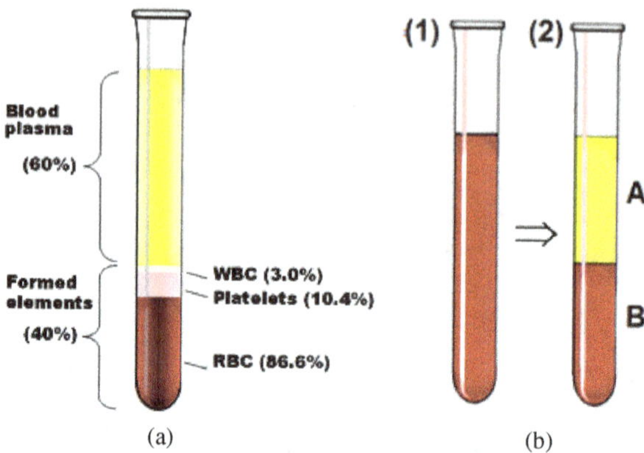

Fig. 6.1. (a) Composition of whole blood: blood plasma and formed elements; and (b) separation of whole blood (in tube 1) into two fractions including A) blood plasma (upper part in tube 2) and B) formed elements (lower part in tube 2).

Blood coagulation involves platelet activation and coagulation cascade. When the platelets encounter the break situation of the blood vessel, external molecules touching the platelets trigger platelet activation; it is followed by the coagulation cascade, which is a complicated step-by-step blood clotting process. Several proteins (fibrinogen, tissue factor, calcium, etc.) and molecules, called coagulation factors, play important roles in the coagulation cascade. Formed elements and blood plasma involve in different ways in contributing to blood coagulation during hemorrhage; they can be separated for the blood droplet tests to single out the direct relevance of CAP to the coagulation cascade in the treatment. The separation is done via precipitation process by sitting the anti-coagulated blood sample in a test tube for 3–5 min at room temperature. As depicted in Fig. 6.1b, the whole blood sample in tube 1 is separated into two fractions: (A) blood plasma (upper part in tube 2) and (B) formed elements (lower part in tube 2), shown in tube 2.

6.1 Blood Coagulation

Coagulation (also known as clotting) is the process to change blood from a liquid to a gel, forming blood clot to halt blood loss from a damaged blood vessel, followed by repair. It involves both a cellular (platelet) and a protein (coagulation factor) component.

After the endothelium lining the blood vessel is damaged in an injury, coagulation begins almost instantly. Leaking of blood through the endothelium initiates two processes: activation of platelets and exposure of subendothelial tissue factor to blood plasma, which ultimately lead to fibrin formation. Platelets immediately form a plug at the site of injury; this is called primary hemostasis. Secondary hemostasis occurs simultaneously; additional coagulation factors respond in a complex cascade to form fibrin strands, which strengthen the platelet plug.

Overall, the mechanism of coagulation involves activation, aggregation, and adhesion of platelets along with deposition, maturation, and crosslink of fibrin. When the blood vessel endothelium is damaged, the circulating platelets are activated by thrombin to release the contents of stored granules into the blood plasma. The activated platelets change shape from spherical to stellate, and the fibrinogen cross-links with platelets' glycoprotein to aggregate adjacent platelets, which are then bound to the collagen exposed on endothelial cell surfaces. A glycoprotein called von Willebrand factor (vWF), which is found in blood plasma, further strengthens this adhesion by binding collagen to the platelets; this binding activates platelet integrins, which mediate tight binding of platelets to the extracellular matrix and thus adhere this platelet plug (white clot) to the site of injury; this is referred to as "primary hemostasis". Oxidants, produced or released in the vascular lumen trigger several key steps in the coagulation cascade. Moreover, the damaged tissues release thrombokinase/thromboplastin (factor III) to react with prothrombin, which, together with Calcium (IV), forms thrombin to convert soluble fibrinogen into insoluble fibrin to deposit into the platelet plug. Platelet phosphatidylserine (PS) exposure keeps coagulation cascade for continuously forming tenase and prothrombinase complexes to amplify thrombin generation. Thrombin also activates platelets to mediate the formation of covalent bonds, which crosslink the fibrin polymers to form fibrin mesh (red clot) all around the platelet plug to hold it in place; this step is called "secondary hemostasis".

In this process some red and white blood cells are trapped in the mesh which causes the primary hemostasis plug to become harder. The mixture of two clots forms a hemostatic plug which is called "blood clot" with blood cells trapped in it.

6.2 In-Vitro Tests of Air Plasma Blood Coagulation

Small wounds can coagulate naturally; on the other hand, serious trauma needs first aid and medicine to stop bleeding. If bleeding is not controlled swiftly, it could become life threatening. An air plasma spray (APS) is employed to clot blood-droplet samples; this is the first step to explore the efficacy of this plasma coagulator as an advanced first aid. In the following, the relevance of atomic oxygen, delivered by the APS to the sample in the treatment, in speeding up the formation of blood clot is studied.

Blood samples used in tests were mixed with 3.2% sodium citrate solution at 9:1 ratio (in volume). Each sample was set on a glass slide. The sodium citrate solution is a commonly used reagent to prevent premature blood coagulation; it chelates calcium ions to prolong the natural clotting time to more than 25 minutes.

The effects of heat and the atomic oxygen radical on blood clotting were studied and compared by tests on blood droplet samples. A test of heating effect was performed by treating a blood droplet set in a well with a hot airflow of a hair dryer for 16 s; it raised the sample temperature from 20°C to about 61°C. A photo of the sample taken after a hot air treatment is presented in Fig. 6.2a. No noticeable blood clot can be identified.

The atomic oxygen flux carried by the air plasma spray drops rapidly with the distance from the nozzle exit; the total amount of atomic oxygen applied to the blood droplet sample in treatment will be proportional to the total exposure time and inversely proportional to the exposure distance. Hence, the plasma treatments on blood droplet samples at different exposure distances and durations can qualitatively reveal the dependency of the coagulation cascade on the applied atomic oxygen flux. Three samples treated at the same exposure distance of 25 mm for 8, 12, and 16 s, respectively, are presented in Figs. 6.2b–d for comparison.

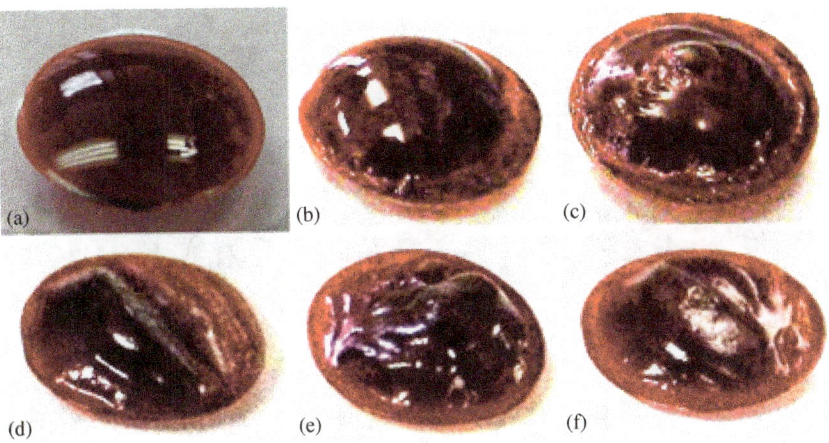

Fig. 6.2. Blood samples treated (a) by a heated airflow for 16 s; by an air plasma spray at a fixed exposure distance of 25 mm with three exposure times of (b) 8 s, (c) 12 s, and (d) 16 s; and by the same air plasma spray with a fixed exposure time of 16 s at two increased exposure distances of (e) 30 mm and (f) 40 mm.

The treated sample temperatures were less than 55°C. A shell, formed on each blood sample surface, can be clearly seen. The photos indicate that the degree of blood clotting increases with the increase of the exposure time from 8 to 16 s. Three samples treated at three different exposure distances of 25, 30, and 40 mm for the same exposure time of 16 s are presented in Figs. 6.2d–f. Again, the treated sample temperatures were less than 55°C. As shown, the degree of blood clotting decreases as the exposure distance increases from 25 to 40 mm. From these results, it is deduced that the degree of blood clotting increases with the increase of the atomic oxygen flux delivered by the APS.

Applying the precipitation process to separate whole blood as shown in Fig. 6.1b, another test was performed on whole blood, blood plasma, and formed elements droplets, to reveal more information of atomic oxygen's role.

Shown in row 1 of Fig. 6.3 are the untreated samples, used as the controls of (a) whole blood, (b) blood plasma, and (c) formed elements droplet-samples. The exposure distance and time were 30 mm and 16 s, respectively. The test results are presented in row 2 of Fig. 6.3

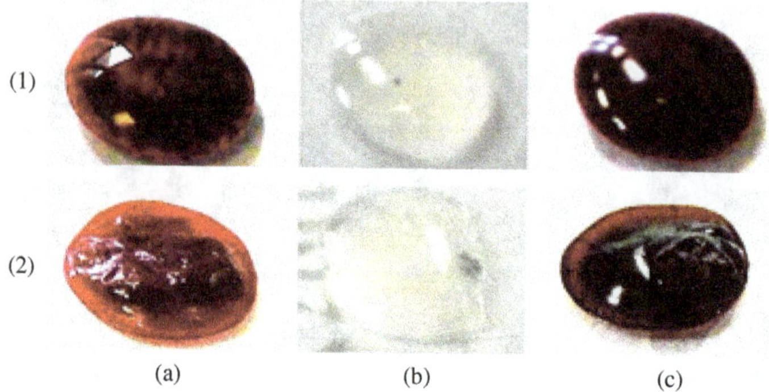

Fig. 6.3.   Untreated controls (row 1) of whole blood (1a), blood plasma (1b) and formed elements (1c) droplet-samples; and the corresponding plasma treated samples (row 2).

for treated (a) whole blood, (b) blood plasma, and (c) formed elements samples. A dark brown shell, manifesting blood clotting, can be seen in the whole blood and formed elements samples but is not observed in the treated blood plasma sample. The treated formed elements sample is darker, dryer, and harder than the treated whole blood sample. The observation evidences that the clotting of the treated samples is ascribed to the activation of formed elements by the reactive species in the plasma effluent.

6.3 Tests on Smeared Blood Samples-Cell Count Dependency

Blood smear technique is applied to spread a drop of blood by a (spreader) glass slide evenly on a glass slide. It keeps every cell from barely touching one another when checked under the microscope for cell count. The smeared blood samples were treated by the APS to explore the changes of RBC and platelet concentrations with the exposure time and distance, through cell counts. Untreated (control) and APS-treated smeared blood samples were prepared for cell staining and microscopy analysis, which identified cell types and performed cell counts. The results of cell counts from samples treated with four different exposure times of 2, 3, 4, and 10 s and

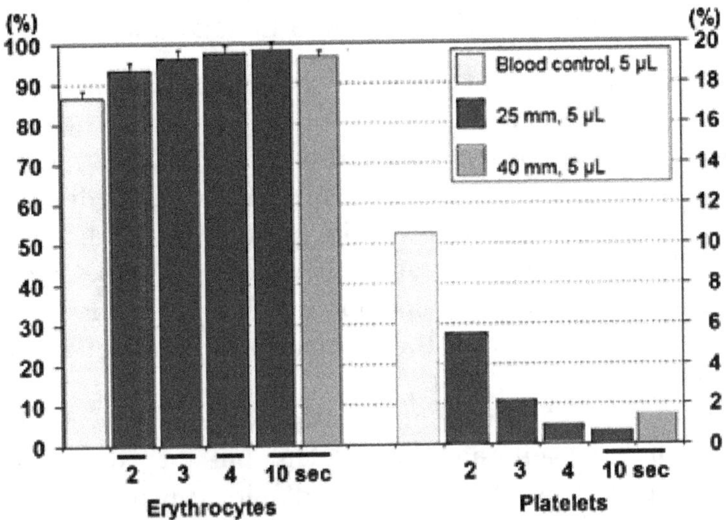

Fig. 6.4. Changes of RBC and platelet concentrations of plasma-treated sample with the increase of the exposure time from 2 to 10 s in the cases of two exposure distances of 25 and 40 mm.

with two different exposure distances of 25 mm and 40 mm are presented in Fig. 6.4.

It is shown that, with a fixed exposure distance of 25 mm, the concentration of RBC/platelets increases/decreases monotonically as the exposure time increases. Cell counts of RBC and platelets with 10 s exposure time at 25 mm and 40 mm exposure distances were compared. As shown, the treatment at longer exposure distance causes less changes on the concentrations of RBC and platelets.

Atomic oxygen, delivered by the APS, induces oxidants (e.g., H_2O_2) in blood plasma. The correlation between the rapid reductions of the platelet count and greatly shortened clot formation time (presented in Section 6.2) of a treated sample suggests that platelets are activated by oxidants to release the contents of stored granules into the blood plasma, which induce coagulation and subsequent platelet clot formation. The activated platelets lose contents and change shape from spherical to stellate; thus the cell count of the (inactivated) platelets in the smeared blood sample decreases with the increase of the treatment time.

6.4 Mechanism of Air Plasma Blood Coagulation

Blood coagulation forms a platelet clot for the primary hemostasis, and a subsequent blood clot for the secondary hemostasis. The overall mechanism of coagulation involves activation, aggregation, and adhesion of platelets along with deposition, maturation, and crosslink of fibrin. Studies have indicated that oxygen free radicals generated chemically, by leukocytes and hemoglobin derived from membrane leakage of erythrocytes, play an important role in the mechanism of platelet activation and aggregation.

The process of primary hemostasis involves mainly activation, aggregation, and adhesion of platelets; but the coagulation of blood droplet does not implicate adhesion phenomenon. The blood droplet tests observe the efficacy of ROS, delivered by CAP, on promoting platelet activation and aggregation, which enhance the release of C_a^{2+} ions, by the activated and aggregated platelets, to overcome being chelated by the anticoagulant. This ROS effect is revealed by the platelet plug formation speed. Calcium (IV) also plays a significant role in forming thrombin, leading to the formation of blood clot in the secondary hemostasis process. Agonist-initiated platelet phosphatidylserine (PS) exposure is regulated by mechanisms different from those of other platelet responses, such as platelet aggregation and granule release. Increased C_a^{2+} level is a key signal initiating PS exposure. It leads PS, which had been limited to the platelet membrane inner leaflet, to rapidly equilibrate between the inner and outer leaflets of the platelet membrane, exposing PS to the plasma milieu. PS exposure amplifies thrombin generation by facilitating assembly of the tenase and prothrombinase complexes.

The results of the tests on blood droplet samples presented in Figs. 6.2d to f and of the microscopy analysis of smeared blood samples presented in Fig. 6.4 show that the blood clotting decreases and the RBC/platelet count decreases/increases with an increase of the exposure distance from 25 mm to 40 mm. Likewise, as shown in Fig. 3.1c, atomic oxygen carried by the APS decreases with increasing exposure distance.

Moreover, in the sequence of tests with increasing plasma treatment time, the correlation between the rising coagulation appearing on the surface of the blood droplet sample, as shown in Figs. 6.2b–d, and the increasing/decreasing of RBC/platelet counts, as shown in Fig. 6.4, was observed. These observations suggest that atomic oxygen carried by the plasma effluent rapidly promotes platelet activation and aggregation and triggers the coagulation cascade; thus the clotting speed increases with the increase of the amount of atomic oxygen delivered by the plasma.

Cartoon plots presented in Fig. 6.5 incorporate these correlations, i.e., correlating the speedy coagulation, the increasing of RBC, and decreasing of platelet counts, to the increase of atomic oxygen supplied in the treatment, to describe a plasma clotting mechanism in the following.

When interacted with H_2O, atomic oxygen can generate a large amount of reactive oxygen species (free radicals and hydrogen peroxides), which are expected to function similarly to those oxidants, produced or released in the vascular lumen. Several key steps in the coagulation cascade are then triggered by the oxidants. These oxidants stimulate RBC–platelets and WBC interactions (Fig. 6.5a). The interactions influence the concentration of cells suspended in blood (Fig. 6.5b). The observations of the microscope presented in Fig. 6.4 show the increase of RBC concentration in line with the increase of the exposure time. Enhanced adenine nucleotides released by aggregated RBC trigger platelet adherence/agglomeration, where fibrinogen acts as a bridge to link activated platelets

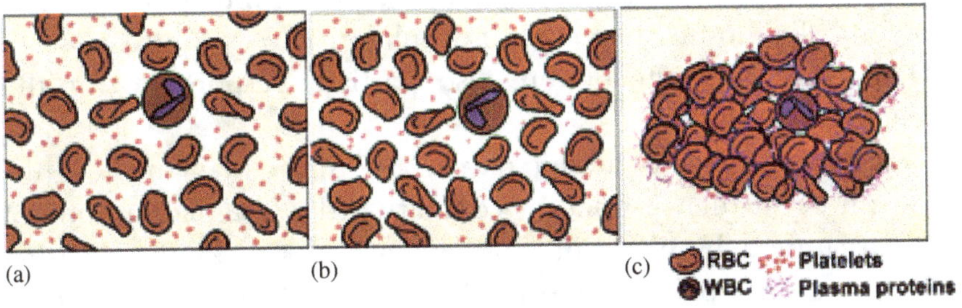

Fig. 6.5. Cartoon plots showing the clotting process in the APS treatment.

together. Oxidants also directly activate and aggregate platelets by inactivating anticoagulation agents, such as endothelial-derived relaxing factor (EDRF) which counteracts platelet aggregation to control clotting. Subsequently, globular complexes (Fig. 6.5a & b) are formed to trap RBC and platelets; the viscosity of blood samples will also be affected by oxidants that presumably contribute to albumin denaturation as well as other proteins found in the blood. Consequently, the blood flowability is decreased and coagulation is rising, leading to the formation of platelet clot (Fig. 6.5c).

NO can lead to anti-inflammation and anti-thrombotic effects which inhibit platelet adhesion, increases blood flowability, and lowers blood pressure. It was demonstrated that NO containing drugs such as an NO donor (e.g., nitroglycerin) or NO inhalant, inhibit platelet agglomeration. NO suppresses platelet agglomeration *in vitro* and *in vivo* via the guanylyl cyclase mechanism. Furthermore, exposure to the inhaled NO significantly decreases platelet agglomeration in rats and in humans accompanied by acute respiratory distress syndrome. This APS operates at a relatively low temperature ($< 55°C$ outside the nozzle). The NO flux in the plasma effluent outside the cap is relatively low and does not significantly quench the effect of atomic oxygen involved in platelet agglomeration and triggering blood coagulation cascade.

It is noted that other reactive oxygen species (ROS), in addition to atomic oxygen, can also create oxidants in the blood to trigger coagulation cascade via the proposed mechanism. A scan of the spectrometer did not reveal oxygen radicals-related spectral peaks other than that around 777.4 nm and 770.68 nm (which is much lower than that of 777.4 nm); however, given the limitations of spectroscopic diagnostics, one could not rule out the existence of other ROS in the APS, which were also partially responsible for the observed coagulation.

The critical role of atomic oxygen/ROS in enhancing coagulation can be further justified by comparing the coagulation time of CAP treatment by different plasma devices. The kINPen Plasma MED is an atmospheric pressure plasma jet (APPJ) using Argon gas as the working gas. It generates ROS only after encountering the ambient air outside the generator. It takes 10 minutes to achieve a complete clot. On the other hand, APS treatment reaches a complete clot in 16 seconds. In addition, the air plasma of a DBD device can also initiate noticeable partial coagulation in 15 seconds. However, because such *in vitro* tests do not involve complete coagulation cascade and processes, the test results cannot really reveal the CAP-efficacy on hemostasis; *in vivo* tests are necessary and are presented in the next chapter.

Problems

P6.1. In a normal situation, blood should be collected in a blue-top tube containing 3.2% buffered sodium citrate. However, the plasma citrate concentration will be too high for those patients who have elevated hematocrits, giving a relatively low amount of plasma for a given whole blood (collection) volume. When the patient has a known hematocrit >55%, the amount of citrate in the collection tube must be decreased according to the formula

$$C = (1.85 \times 10^{-3}) \, (100-\text{Hct}) \, (V_{\text{blood}})$$

where C is the volume of citrate remaining in the tube, Hct is the patient's hematocrit, and V_{blood} is the volume of blood added to the evacuated tube.

Calculate the amount that the citrate in the collection tube should be adjusted to if the Patient hematocrit = 60%.

P6.2. Use blood collected in a blue-top tube containing 3.2% buffered sodium citrate to perform a coagulation test. Set a few blood droplets on a glass slide and measure the (average) natural clotting time. Compare the measurement result with those presented in Fig. 6.2 and summarize the apparent effects of plasma treatment.

P6.3. Place a glass slide containing blood droplet samples, similar to that prepared in problem P6.2, in a glass ware containing concentrated (35% or 59%) hydrogen peroxide liquid (Vaprox), similar to that used in P5.4. The partially covered glass ware is heated at 30°C to vaporize Vaprox. Measure the (average) clotting time of the droplets under the VHP treatment. Repeat the tests with the heating temperature increased in 5°C steps up to 50°C, to plot the clotting time dependency on the VHP flow.

P6.4. If the air plasma generator outlined in problem P2.6 has been set up, one can measure the clotting times of the blood droplet samples treated by the CAAP of the air plasma generator. Set the blood droplets in the glass slides and measure the clotting times of the blood droplet samples under the treatment of the CAAP at 25, 30, and 40 mm exposure distances; and compare the results with the measurements of problem P6.3.

P6.5. The Harboe spectrophotometric assay uses absorbance measurement at 405 nm in a microtiter plate reader to gauge hemoglobin for the quantification of blood coagulation.

The procedure: 1) wash the treated blood sample with PBS (phosphate buffered saline) to separate the unclotted blood portion from the formed blood clot, and pipette it into an Eppendorf tube to be acidified into a 3% acetic acid solution; 2) dissolve the remaining blood clot in separate Eppendorf tube by using 3% acetic acid in PBS, 3) wash the well of dwelling sample with 3% acetic acid to remove any residual hemoglobin and then transfer it to a separate Eppendorf tube; 4) appropriately dilute both the unclotted fraction and dissolved blood clot for absorbance readings; 5) record readings by a Synergy HT micro-plate reader; and 6) compare these results with those from an untreated whole-blood sample as well as a control blood clot, respectively.

(1) Explain that a blood clotting index (BCI) can be defined to be

$$BCI(\%) = \frac{\text{Absorbance of acidified dissolved blood clot}}{\text{Absorbance of acidified dissolved control blood clot}} \times 100\%$$

$$= \left[1 - \frac{\text{Absorbance of acidified unclotted blood portion}}{\text{Absorbance of acidified whole blood sample}} \times 100\% \right]$$

(2)   Estimate the likely deviation,

(3)   Check BCI for the results of P6-3 and P6-4.

P6.6.   Blood films are made by placing a drop of blood on one end of a glass slide, and using a spreader slide to disperse the blood over the slide's length. As the spreader slide draws the blood forward, the front edge emerges to be a monolayer, in which the cells are spaced far enough apart to be counted and differentiated. Apply the approach described in problem P6-3 to perform VHP treatment on blood film samples; and implement cell staining and microscopy analysis on untreated (control) and VHP-treated smeared blood samples. Plot the time dependence of the cell counts (normalized to those from the untreated sample) from the monolayers of the samples (treated at four different exposure times of 5, 10, 15, and 20 s).

P6.7.   Repeat problem P6.6 and replace VHP treatment by CAAP treatment at 25 and 40 mm exposure distances; and change the exposure times to 4, 8, 12, and 16 s.

Chapter 7

Air Plasma Bleeding Control Study with Animal Models

7.1 Background

Injury results from a wide variety of causes, including accidents or intentional harm, and in a wide variety of locations, such as home, workplace, on the road, and battlefield. According to the Department of Homeland Security, it can take less than five minutes for someone to bleed to death. Uncontrolled bleeding is the number one cause of preventable death from trauma. Successful control of bleeding under extreme circumstances will result in saving lives.

Basic lifesaving steps for the medic include clearing the airway/restoring breathing, stopping the bleeding, protecting the wound, and treating/preventing shock. Certain types of wounds and burns require special precautions and procedures when applying these measures. When properly applied, these techniques can save lives. It is important that as many people as possible survive their injuries.

First aid is appropriate for external bleeding. However, EMS and other forms of help often take time to arrive on scene; and life threatening bleeding will need an advanced first aid to control it swiftly.

7.2 Bleeding

Blood flows in blood vessels which include arteries, arterioles, capillaries, veins and venules. Arteries carry blood away from the heart; an arteriole is the smallest branch of an artery. Capillaries are tiny blood vessels found throughout the body, where gases, nutrients, and waste products are exchanged. Veins carry blood from capillaries back to the heart; a venule is the smallest branch of a vein.

When there is a cut, some of the blood vessels are broken to cause bleeding, which is the loss of blood. Bleeding may be inside the body (internally) or outside the body (externally). Here, we consider bleeding occurs outside the body when blood moves through a break.

Sometimes, relatively minor injuries can bleed a lot. An example is a scalp wound. Another case where an injured person can bleed a lot is if this person takes blood-thinning medication or has a bleeding disorder such as hemophilia. Controlling the flow of blood following vascular injury, in particular in a serious situation, is paramount to continued survival. The most important step for external bleeding is to apply direct pressure. This will stop most external bleeding.

7.3 Hemostasis

Hemostasis is a process which stops bleeding and keeps blood within a damaged blood vessel. This involves coagulation which transforms blood from a liquid to a gel forming blood clot. Hemostasis comprises four major events that occur in a set order following the loss of vascular integrity:

1) Vascular constriction limits the flow of blood to the area of injury.
2) Platelets are activated by thrombin and change their shape. The protein fibrinogen aggregates platelets and stimulates platelet clumping by binding to collagen at the site of injury, forming a temporary, loose platelet plug.
3) A fibrin mesh (also called the clot) forms and entraps the plug.
4) Clot is dissolved, through the action of plasmin, for resuming normal blood flow following tissue repair.

When endothelial injury of the blood vessel occurs, platelets, which come across the injured endothelium cells, activate, change shape, release granules, and aggregate, and ultimately become "sticky". Platelets express certain receptors, some of which are used for the adhesion of platelets to collagen. The endothelial cells secrete von Willebrand factor which initiates the maintenance of hemostasis by binding collagen to the platelets' glycoprotein to further strengthen this adhesion. Concurrently, oxidants, produced or released in the vascular lumen, trigger several key steps in a coagulation cascade. Once the platelet plug has been formed, the clotting factors activate the "coagulation cascade" to form thrombin, which transforms inactive fibrinogen plasma protein to a fibrin mesh covering all around the platelet plug to hold it in place. In sum, hemostasis proceeds at the steps: 1) vasoconstriction, 2) temporary blockage of a break by a platelet plug (white clot), and 3) blood coagulation, or formation of a fibrin clot (red clot) to reinforce platelet plug. These processes seal the hole until tissues are repaired to proceed the dissolve of the clot.

7.4 Clotting

When tissue is first wounded, blood comes in contact with collagen, triggering blood platelets to expose sticky glycoproteins on their cell membranes that allow them to aggregate at and to stick to the injured site, forming a mass. The activated platelets change into an amorphous shape, more suitable for clotting, and release chemical signals to promote clotting. This results in the activation of fibrin, which forms a mesh and acts as "glue" to bind platelets to each other. Fibrin and fibronectin cross-link together and form a plug that traps proteins and particles and prevents further blood loss. This fibrin-fibronectin plug is also the main structural support for the wound until collagen is deposited. Migratory cells use this plug as a matrix to crawl across, and platelets adhere to it and secrete factors. The clot is eventually lysed and replaced with granulation tissue and then later with collagen.

Collagen deposition is important because it increases the strength of the wound; before it is laid down, the only thing holding the wound closed is the fibrin-fibronectin clot, which does not provide much resistance to traumatic injury.

In Chapter 6, in vitro tests have demonstrated the efficacy of an APS on clotting blood-droplet samples. The atomic oxygen delivered by the APS essentially activates platelets and triggers coagulation cascade. The clotting time is shortened by two orders of magnitude. In the following, in vivo tests applying the same APS to speed up hemostasis are described.

7.5 Experimental Arrangement

Using pigs as animal models and air plasma spray (APS) in the treatments, experiments were conducted to stop bleeding from four types of wounds, including a straight cut, a cross cut, a hole into an ear saphenous vein, and a cut to an ear artery, in an effort to develop an advanced first aid tool for bleeding control. The observations of wound healing after plasma treatments will be presented in the next chapter.

Two 3 month-old male pigs weighing around 25 kg were used in the experiments of performing first three types of cut wounds; the wound introduced on one pig was treated by the APS; the other pig was an untreated control whose wound bleeding was allowed to stop by itself. Three 6 month-old male pigs, weighing around 40 kg, were used in the artery cut experiment, and in post-operative observation of the healing of the cross cut wound. The left ears of the pigs were the control group and the right ears were the experimental group. The bleeding time of a similar wound on the untreated control was recorded to be the natural clotting (bleeding) time for comparison with that of the corresponding treated wound. The cross cuts in the ham area were also performed on the three large pigs for post-operative observation of the effect of APS on wound healing.

Each pig was first injected with calmative-Stresnil and fastened on a table. The pig was then anesthetized with Isoflurance-Fluothane which kept it in a narcotized state. In the first group of two younger pigs, tests were conducted in the sequence of a straight cut first, then a cross cut, and finally a hole in an ear saphenous vein, in the order of increasing difficulty of bleeding control. Enough time between the tests was given for the pigs to recover from the bleeding. In the next group of three older pigs, tests with a cut to an ear artery, were first conducted; after the artery cut tests, tests of cross cuts in the ham area were performed again for the observation of the effect of APS on the healing of these two different types of wounds, ear artery cut and ham cross cut. After the experiments, the pigs, in the second group, were put into stainless experiment cages for postoperative observation of recovery, which is presented in the next chapter. The stainless cage prevents the pig from scratching an itchy part of the wounds against the wall during the recovery period.

7.6 Surface Wounds

Bleeding is caused by the break of capillaries underneath the skin in the injured area. This is a type of small wound, but bleeding can last long and needs attention.

7.6.1 *Test 1 — straight cut*

A scalpel was used to make a straight cut to each pig in the same ham area. The size of each cut was about 10 mm in length and 5 mm in depth. One cut was not treated. Presented in Fig. 7.1a is a photo of this untreated cut taken 190 s later when the hemorrhage stopped naturally. This time is recorded as the total bleeding time.

The straight cut on the second pig was treated by APS at an exposure distance of 25 mm. The bleeding stopped completely after 18 s of continuous plasma treatment, as demonstrated in Fig. 7.1b. The treated cut appears to be covered only by small amount coagulated blood. This is because in the treatment, the airflow from the spray blew away the remaining blood which did not clot in time. Two similar tests with the exposure distance increased to 30 and 40 mm were also performed. The coagulation times in two cases were measured to be 17 and 21 s, respectively.

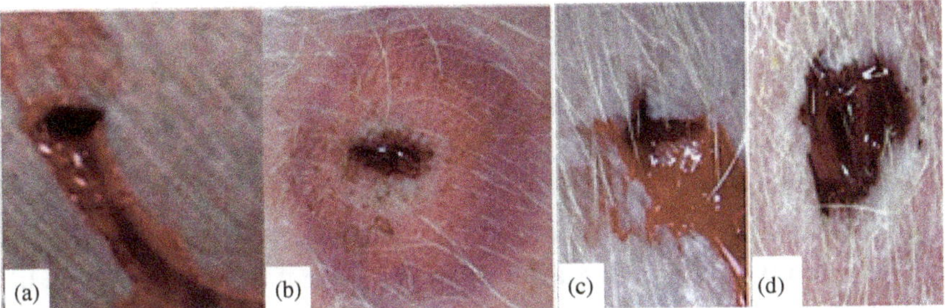

Fig. 7.1. Photos of straight cuts (a and b) and cross cuts (c and d) taken (a) after more than 3 minutes waiting time for bleeding to stop naturally, (b) after bleeding stopped by a plasma treatment of 18 s, (c) after 4 minutes waiting time for bleeding to stop naturally, and (d) after the bleeding stopped by 13 s plasma treatment.

7.6.2 *Test 2 — cross cut*

The next test performed was a cross cut, which consisted of two straight cuts cross to each other, to each pig in the same ham area. Again, the size of each cut was about 10 mm in length and 5 mm in depth and the APS exposure distance was 25 mm. The hemorrhage from the untreated cross cut lasted for more than 4 minutes. A photo of the cut taken after the bleeding stopped naturally is presented in Fig. 7.1c. On the other hand, the cross cut wound on the other pig was treated by APS continuously and the bleeding was stopped in 13 s. A photo of this cross cut sealed by coagulated blood, with the aid of the plasma treatment, is presented in Fig. 7.1d. The treatments needed for the other two exposure distances of 30 and 40 mm were measured to be 17 and 22 s, respectively.

7.7 Vessel Wounds

This involves injury to an artery or a vein. The primary consideration in the treatment of blood vessel wounds is hemostasis and this must be thorough and absolute, as without it the danger of wound infection and subsequent grave danger to the patient's life are inevitable.

First Aid Treatment is essential and must be prompt and efficient; such treatment consists, in most cases, in the application of local pressure to the wound by means of a surgically sterile dressing and this usually suffices to stop the bleeding. In case of damage to large arteries or where no sterile dressing is available, it is best to apply some form of tourniquet to the limb immediately above the wound, care being taken that this is applied only just so tightly as to arrest the bleeding, and that the tourniquet is loosened every 20 minutes to reestablish arterial flow. It enables the peripheral circulation to reestablish through collateral vessels to avert gangrene beyond the site of the wound; this is especially the case in the lower limb.

In the tests, small blood vessels in the ears are adopted. Bleeding of those blood vessels can stop naturally, so that no stitches would be needed. The bleeding time is determined directly from the test, which makes it easy to compare the bleeding times of the untreated controls and treated wounds. A significant reduction of the bleeding time demonstrates the efficacy of plasma treatment.

7.7.1 *Test 3 — hole in a saphenous vein*

A saphenous vein in a pig ear was first identified as shown in Fig. 7.2a; a needle and forceps were then used to punch a hole in this vein. When blood flow started, it was treated immediately by APS with an exposure distance of 25 mm. The bleeding stopped in 15 s as demonstrated in Fig. 7.2b. In the untreated control, the bleeding time of the other pig was measured to be about 88 s.

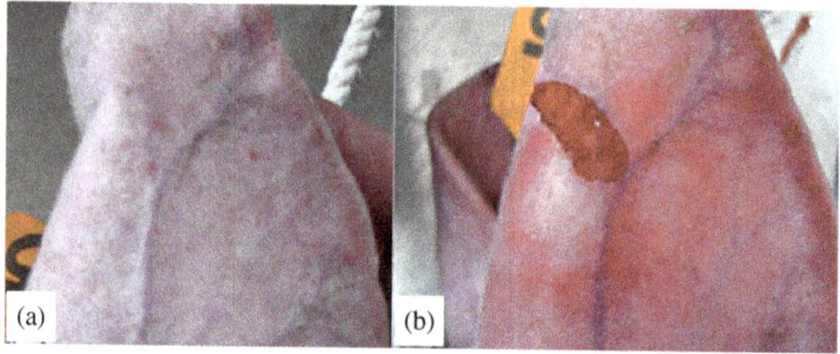

Fig. 7.2.   (a) Photo showing a saphenous vein in an ear of a pig, in which a hole is to be punched and (b) photo of the punched saphenous vein after 15 s plasma treatment to block the hole from bleeding.

7.7.2 *Test 4 — a cut to an artery*

Before cutting an artery, the ear was tied with a tourniquet to slow down the blood flow. A scalpel was then used to cut the ear small artery as shown in Fig. 7.3a. The needed plasma treatment time and the natural coagulation time were measured for comparison. In the untreated case, leaving the bleeding unattended as shown in Fig. 7.3b, it took 1 minute for the bleeding to stop naturally. However, the tourniquet could not be removed immediately after the bleeding was stopped.

In the plasma treatment as shown in Fig. 7.3c, an intermittent approach, with plasma 2-s on/4-s off alternately, was adopted.

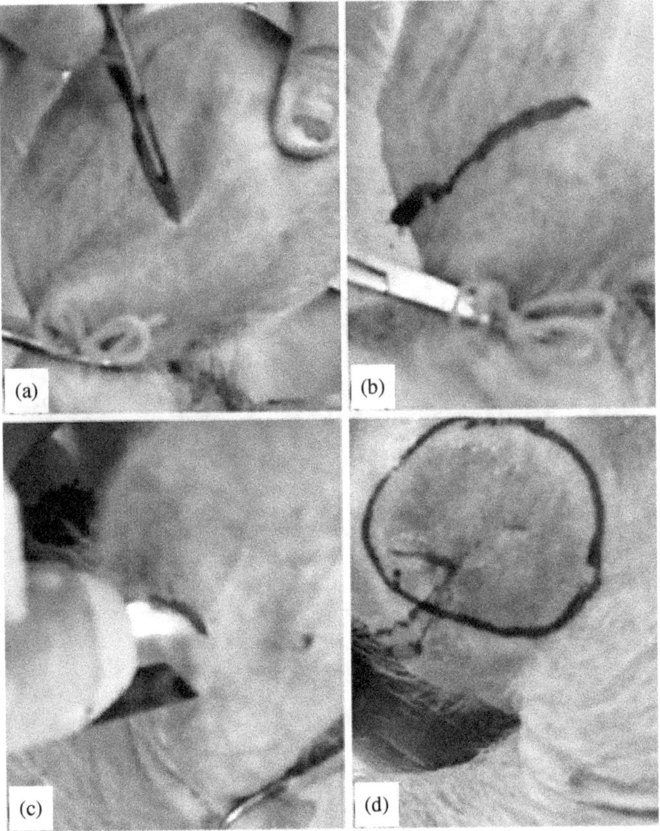

Fig. 7.3. Photos showing (a) cutting an artery and (b) letting the bleeding stop naturally; and (c) with plasma treatment and (d) stop the bleeding in half time.

Bleeding was stopped after 6 runs of plasma on-off treatment as shown in Fig. 7.3d. The total treatment time was about 35 s, about half of the natural clotting time; however, the total plasma exposure time was only 12 s. More significantly, the tourniquet could be removed right after the bleeding was stopped. The plasma treatment also irritated the exposure area around the wound, which is circled in Fig. 7.3d. This irritation disappeared in several days, and no any apparent side effect was observed (see the experimental examination presented in next section).

Adopting an intermittent plasma treatment minimizes the thermal factor in the overall plasma effect on wound bleeding control. However, in practical applications, continuous treatment should be adopted, in particular, in an emergency situation, to further reduce the bleeding time.

7.8 Discussion

The plasma treatment time and the natural time (control) for the bleeding to stop from each type of wound cut in the tests are recorded in Table 7.1. A comparison of these experimental results clearly shows that this plasma spray rapidly clots blood to stop bleeding.

Table 7.1   Bleeding control time of air plasma treatment.

| | | Treatment | | |
| | | | | |
Wound	Plasma (continuous)	Plasma (intermittent) 2-s on/4-s off On time	Plasma (intermittent) 2-s on/2-s off On time	Natural (control)
Straight cut (1 cm × 0.5 cm)	18 s	8 s	10 s	190 s
Cross cut (1 cm × 1 cm × 0.5 cm)	13 s	8 s	10 s	240 s
A hole in an ear's saphenous vein	15 s			88 s
Artery cut in ear (tourniquet aided)		12 s		60 s

In Sec. 6.4, atomic oxygen produced in the plasma effluent is suggested to be the catalyst in the coagulation cascade.

In order for hemostasis to occur, platelets must adhere to exposed collagen, release the contents of their granules, and aggregate. The adhesion of platelets to the collagen exposed on endothelial cell surfaces is mediated by the von Willebrand factor (vWF). The function of the vWF is to act as a bridge between a specific glycoprotein complex on the surface of platelets (GPIb-GPIX-GPV) and collagen fibrils.

Activation of platelets is required for their consequent aggregation to a platelet plug. Tissue factor is a cofactor in the factor VIIa-catalyzed activation of factor X, which is then cleaved to Factor Xa by Factor VIIa. Active factor Xa hydrolyzes and activates prothrombin to thrombin, which then converts fibrinogen to fibrin. The formation of complex between factor VIIa and tissue factor is a principal step in the overall clotting cascade.

Tissue factor pathway inhibitor (TFPI) is a serine-protease inhibitor, which modulates coagulation tissue factor-dependent. Studies show that ROS promote a pro-coagulant state via tissue factor (TF) expression. It is by altering TFPI structure that a pro-coagulant state in endothelial cells is induced. Nitric oxide (NO) produces vasodilatation, reduces platelet aggregation to the endothelium, prevents leukocyte adhesion and infiltration into the vessel. ROS can impair the NO production by the endothelium to reduce its vasodilator activity. Plasmin(ogen) regulates coagulation; but reactive oxygen species (ROS) can regulate plasminogen-induced IL-1β and TNF-α production in microglia to degrade plasmin in anticoagulation.

Studies also show that platelets are a prime target for oxidants (ROS) produced or released in the vascular lumen and, at the same time, they are also capable of endogenous generation of oxidants. ROS affects cell metabolism, the induction of host defense genes, and mobilization of ion transport systems. This implicates them in control of cellular function. Platelets involved in wound repair and blood homeostasis release ROS to recruit additional platelets to sites of injury. These also provide a link to the adaptive immune system via the recruitment of leukocytes.

In summary, when atomic oxygen interacts with H_2O, ROS are produced to promote a pro-coagulant state, to reduce NO's vasodilator activity, and to degrade plasmin in anticoagulation. The viscosity of blood is affected by oxidants that presumably contribute to the denaturation of albumin and other proteins found in the blood. Consequently, the blood flowability is decreased and coagulation is rising. These oxidants also target platelets, in the same way as those produced or released in the vascular lumen, to affect several key steps of platelet function, such as enhancing platelet aggregation and thrombin formation. H_2O_2 stimulates phospholipase A2 enzyme to amplify platelet response to collagen and acts as second messenger by activating arachidonic acid metabolism and phospholipase C (PLC) pathway. It rapidly advances platelet aggregation which sets off an increase in intracellular C_a^{2+} ions, combining with prothrombin to form thrombin.

The soluble fibrinogen is then converted by the induced thrombin into insoluble fibrin strands to form fibrin gel, which holds activated platelets to form homeostatic plug for coagulation.

The test results show that the APS is effectual to trigger various reactive processes in the blood coagulation cascade, crosslinking various blood polymers and activating platelets, which prepares them for clot formation and speeds up hemostasis.

Moreover, the APS is able to kill bacteria by protein degradation in surface structures and damaging microbial DNA without being harmful to human tissues. Reactive oxygen species (ROS) and other components of plasma affect eukaryotic cells, especially on keratinocytes in terms of viability, proliferation, DNA, adhesion molecules and angiogenesis.

Thus, the APS can also support wound healing, as demonstrated in the next chapter, by its antiseptic effects, by stimulation of proliferation and migration of wound relating skin cells, by activation or inhibition of integrin receptors on the cell surface or by its pro-angiogenic effect.

Problems

P7.1. Wounds of the brachial artery, femoral artery, and popliteal artery cause severe bleeding, which needs to be controlled quickly; otherwise, the bleeding can contribute to shock, circulatory disruption, or more serious health consequences such as damage to tissues and major organs, which can lead to death. Describe the emergency steps to control the bleeding before receiving professional medical help.

P7.2. The duration of acute pain from trauma is often wrongly assumed to be limited to the healing time of the damage. However, any factor contributing to the continuing barrage of nociceptive signals may increase the risk of long-term hypersensitivity which causes chronic pain.

Explain why the CAAP treated trauma has a lower risk of long-term hypersensitivity.

P7.3. When the wound of an injury starts to bleed, 1) the injured blood vessels become narrower to reduce the flow of blood to the injured tissue, limiting the loss of blood; 2) blood platelets in the bloodstream, known as thrombocytes, attach to the damaged area of the blood vessel and clump together to reduce the bleeding; 3) the body then activates a number of substances in the blood and the tissue. These substances solidify the clump by forming a special protein and fix the clump at the wound. These substances are called clotting factors or coagulation factors. There are 13 clotting factors in human blood and tissues. Most of them are made in the liver. The liver needs vitamin K to make some of these clotting factors. Our bodies cannot make their own vitamin K, so we have to get it in the diet. List foods which can provide vitamin K.

P7.4. Placing pressure on the wound to constrict the blood vessels manually, helps to stem blood flow. This is a step for bleeding control of traumatic injury. Arterial bleeding requires more specific pressure on the bleeding vessel than the generalized pressure used for venous-type bleeding. This may require finger-tip pressure at the point where the bleeding is coming from, rather than a generalized pressure on the wound itself. This is due to the higher pressure of the arterial system. What are additional attentions needed in performing treatment and how can the CAAP treatment complement the requirements?

Chapter 8

Wound Healing

8.1 Healing Process

In Mammalian skin, the outermost layer is the epidermis which has protection formation and waterproof property; the inner layer, the Dermis, provides a location for the appendages of the skin. The hair follicles sweat glands, sebaceous glands, lymphatic vessels, blood vessels etc. are contained inside the dermis. The definition of a wound is a break in the epithelial integrity of the skin. The disruption could be deeper, extending to the dermis, muscle or even the bone. The entire wound healing process is a complex and dynamic process of restoring cellular structures and tissue layers in which the damaged skin is being repaired. The physiologic process of wound healing goes through two phases, the early and the cellular phases, which sequentially cover hemostasis, inflammatory response, proliferating and remodeling.

The early phase, which begins immediately following skin injury, involves cascading molecular and cellular events leading to hemostasis and formation of an early, makeshift extracellular matrix that provides structural staging for cellular attachment and subsequent cellular proliferation.

The cellular phase involves several types of cells working together to mount an inflammatory response, synthesize granulation tissue, and restore the epithelial layer.

After a wound starts to bleed, the endothelium of the damaged vasculature stimulate platelets to release certain factors, which result in vasoconstriction and the aggregation of platelets. Thrombin, growth factors, prostaglandins, and other cytokines are released at the site of the injury to initiate coagulation cascade; platelets then reach homeostasis by forming a fibrin clot. The alpha granules of the platelets contain growth factors, and those proteins start the wound healing cascade by attracting and activating fibroblasts, endothelial cells and macrophages. The platelets trapped in the clot are essential for homeostasis as well as for a normal inflammatory response.

In the inflammatory stage of healing, polymorph nuclei arrive at the wound site about one hour after injury, and adhere themselves to the endothelial cell walls of the damaged vasculature to become the predominant cells. These phagocytes provide protection against infection; they release enzymes, free radicals, and reactive oxygen species to kill bacteria; they also phagocytize dead tissue and any foreign material present.

Those neutrophils also produce inflammatory mediators that activate and recruit fibroblasts and epithelial cells to the injury site. When neutrophils are depleted they are replaced by macrophages, which help to rid the wound of devitalized tissue and produce elastase and collagenase. Macrophages also prompt an ending of the inflammatory phase and the beginning of the proliferative phase of wound healing. Factors are released to cause the migration and division of cells involved in the proliferating stage, which is characterized by the replacement of the provisional fibrin matrix with newly formed granulation tissue. This rudimentary tissue contains new blood vessels, fibroblasts, endothelial cells, etc.

The remodeling stage starts concurrently with the development of granulation tissue. In the remodeling phase, collagen matrix is remodeled and collapsed along tension lines and cells. The vessel cells and muscle tissues grow toward the steady state under the collagen matrix. Collagen degrades to scar which is then removed because of apoptosis. This explains why wound healing takes time.

The wound healing time varies with the location, age, degree of wound, diet, etc., but the timing of the healing process is important to wound healing. Critically, the timing of wound re-epithelialization can decide the outcome of the healing. If the re-epithelization of tissue over a denuded area is slow, a scar will form over many weeks, or months; on the other hand, if the re-epithelization of a wounded area is fast, the healing will result in regeneration.

8.2 Immune System

The immune system plays a crucial role in the process of wound healing, which progresses sequentially through hemostasis, inflammation, proliferation and remodeling phases. As bleeding occurs, platelets are activated and aggregate at the damaged site to initiate clot formation, which stops blood loss and creates a temporary covering scab to provide protection from the external environment. Platelets also secrete factors that recruit other immune cells and initiate the inflammatory phase.

Neutrophils of Immune cells first enter the damaged site to remove foreign material and bacteria from the wound. Monocytes are the next wave of immune cells, which differentiate into cells of macrophages. These cells use "eating" capacity to clean the debris at the damaged site, laying the foundations for tissue repair in the proliferative phase.

Macrophages also help promote the reconstruction of the tissue's extracellular matrix, a biological scaffold for the cells that will form new tissue, by producing fibroblast growth factor to promote the growth of fibroblasts. These cells produce the precursor components of the extracellular matrix to provide structure to the tissue as a temporary matrix, which is replaced later by a stronger matrix in the remodeling phase. Once this matrix is repaired, macrophages can promote the growth of skin cells that fill the previously wounded area. Fibroblasts within this site also differentiate into myofibroblasts that resemble muscle cells. These myofibroblasts make a permanent closure of the wound, preventing exposure to the external environment.

If the wound is deep enough to disrupt the vasculature, new blood vessel formation is also guided by macrophages, which are secreting proteins to recruit the cells that form blood vessels, and then pattern their growth.

Bacteria increases the inflammatory response; excess inflammatory mediators can cause tissue damage and may sometimes prolong the inflammatory phase of healing. CAP treatment reduces the levels of bacteria present in the wound. Obviously, lessening bacterial burden will lower the degrees of inflammatory mediators, which result in shortening the inflammatory phase of healing.

The impacts of APS plasma treatment on the exposure area around and on the healing time of the wound, as well as on the timing of the healing process, are examined in the following experiments.

8.3 Post-Operative Observation of Wound Healing After APS Plasma Treatment

Post-operative observation is feasible in an in-vivo situation. It helps to understand the plasma effluent effect upon the skin tissue surrounding the treated wound and upon the progress of wound recovery. After the experiments, the pigs were housed in stainless experiment cages to keep them from rubbing their wounds. The progress of recovery was observed by recording the changes of the controls and treated wounds every two days for 14 days.

The healing process of the artery cut wound and its surrounding irritation in the circled area of Fig. 7.3d can be seen in a sequence of 6 photos, taken every other day at day 2 to day 12 after the plasma treatment, presented in Figs. 8.1a to f. The scab appears first at the cut (Fig. 8.1a); then, the surrounding irritated area turns to dark brown color (Fig. 8.1b); the scab starts peeling in the 8th day (Fig. 8.1d) but the dark color in the irritated area starts diminishing in the 6th day (Fig. 8.1c). The dark color disappears completely in the 10th day (Fig. 8.1e) and a complete healing of the cut is observed in the 12th day (Fig. 8.1f). No apparent side effect on the irritated area can be seen.

To show that there was no scab formed in the area irritated by the plasma exposure and no blemish left on the skin, a ring shape irritation was introduced in a wound-free ham area, by continuous plasma exposure at a distance of 25 mm for 10 s, for the observation.

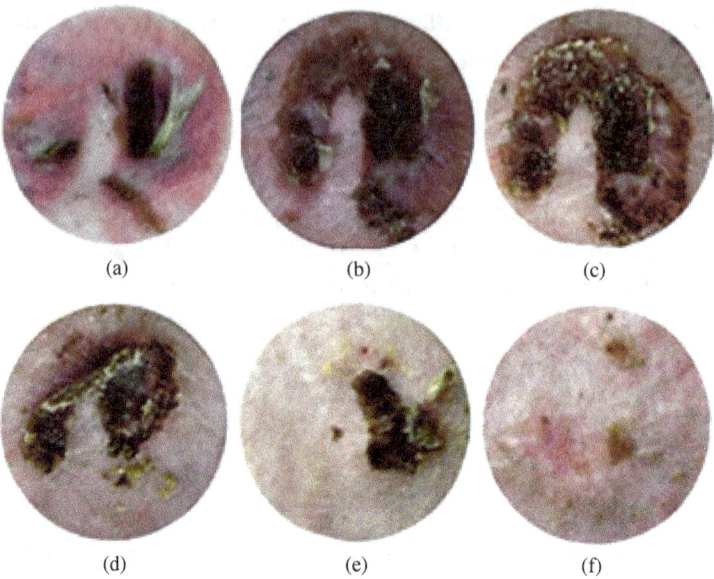

Fig. 8.1.   Recovery of an artery cut and the surrounding irritated area shown in Fig. 7.3d, after the plasma treatment.

The photo records taken from day 0 to day 14 are presented in Fig. 8.2. As shown, in the healing process, the red ring turns to a dark brown ring; the color becomes darker while the dark brown area is diminishing. It disappears completely in the 14th day. In the entire healing process, no scab is formed.

Cross cut wounds were introduced in the ham area of three pigs which were 6-months-old and had a weight of about 40 kg; one was untreated as a control and the other two were treated by the APS with two different intermittent exposure approaches, which applied two different sets of running parameters $(T_E, T_p, N, D) = (2, 4, 4, 30)$ and $(2, 2, 5, 30)$ for a comparison of the outcomes, where T_E, T_p, N, and D represent the plasma-on time in seconds at each run, plasma-off (pause) time in seconds between two runs, number of runs in a treatment, and exposure distance in mm, respectively. The respective photos recording the healing progress of the control and of two treated cuts are presented in three rows in Fig. 8.3, for comparison.

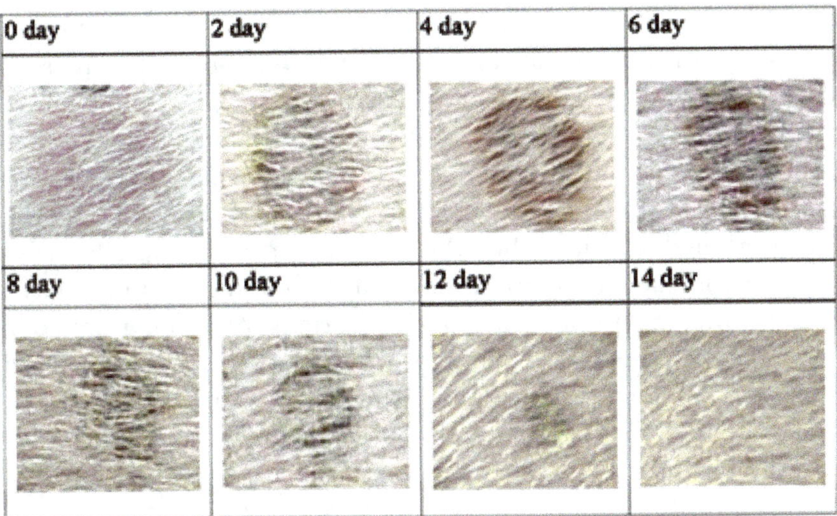

Fig. 8.2. Recovery of an irritated area impinged by plasma spray for 10 s at an exposure distance of 25 mm.

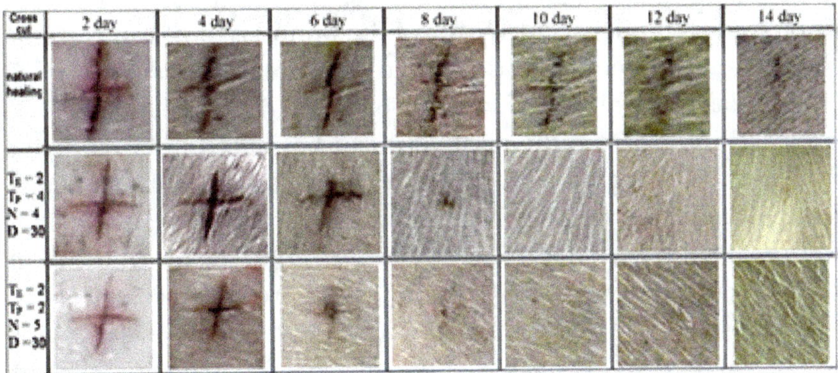

Fig. 8.3. Comparison of the progress of the recovery of untreated (row 1) and treated (rows 2 and 3) cross cuts.

The progress of the control (untreated cross cut) presented in row 1 of Fig. 8.3 indicates that the scab starts peeling in the 8th day. The scab is diminishing in time; however, a small piece of the crust still remains in the wound area in the 14th observation day.

The cut in row 2 of Fig. 8.3 was treated with 8 s ($T_E \times N$) plasma exposure; it is found that the healing time is shortened. The scab starts peeling in the 6th day, and the crust disappears completely in the 10th day. As the plasma exposure time is increased to 10 s, the healing progress of the cut in row 3 of Fig. 8.3 becomes even faster; the starting peeling time of the scab is reduced to 4 days. Moreover, it is shortened to 8 days when the crust disappears completely.

The comparison demonstrates that the plasma treatment has a positive impact on wound healing; it shortens the healing time of a cross cut wound to about half. Moreover, there is no scar is formed at each treated wound site. It indicates that plasma treatment also speeds up the epithelization of the wounded area, facilitating regeneration healing.

8.4 A Plausible Mechanism

Hypoxia acts a key factor to stimulate tissue repair by creating an oxygen gradient from the hypoxic tissue of the wound to the nearby unbroken tissue. The central area of the wound is most hypoxic, and the oxygen gradient increase toward the uninjured tissue progressively. With the supply of atomic oxygen from APS, more ROS provided in situ from the treatment can be shared in other action such as stimulating superoxide dismutase enzymes (SODs), enhancing cell metabolism and raising tissue oxygen tension effect in the wound healing. Atomic oxygen also provides the oxygen in the blood by reaction of catalase which plays a protective role avoiding cells damaged by H_2O_2. More significantly, atomic oxygen together with generated ROS (such as *OH and H_2O_2), via interacting with H_2O, work for disinfection concurrently with blood coagulation in the wound treatment. Moreover, the redness, shown in Fig. 7.3d, indicates that there is extra blood rushes to the skin's surface around the wound in the plasma treatment to encourage healing, i.e., the dilation of local capillaries helps to speed immune cells to enter the damaged site. Thus the period of the inflammatory phase in wound healing is shortened. Cells are aging via accumulating free radical and oxidative damage over time; in order to maintain the normal function, skin tissue is metabolized when it is aging. Because the treated area accumulates sufficient atomic oxygen, this area is aging faster than surrounding tissue. The skin tissue increases the metabolism to speed up the generation of new tissue. The new skin tissue grows under the aging tissue, and replaces the position after aging tissue is peeled.

8.5 Discussion

Human skin consists of two tissue layers, epidermis and thick dermis underlay. The epidermis is a stratified squamous epithelium, which is composed of proliferating and differentiated keratinocytes. The dermis is a collagen-rich connective tissue, which provides support and nourishment.

Dermal fibroblasts play a pivotal role in maintaining the skin by synthesis and deposition of various extracellular matrix (ECM) proteins. In the process of wound healing, dermal fibroblasts synthesize relevant cytokines such as keratinocyte growth factor. Proliferating fibroblasts form a provisional extracellular matrix, in fibroplasia and granulation tissue formation, by generating and depositing collagen and fibronectin. After a few days, a portion of fibroblasts transform into myofibroblasts, which provide physiological reconstruction of connective tissue by resembling smooth muscle cells in their capacity for generating strong contractile forces.

The contractile activity by mature myofibroblasts is beneficial for tissue remodeling and after wound closure, myofibroblasts regularly undergo apoptosis. On the other hand, the persistence of active contractile myofibroblasts can become a detrimental pathological factor in the development of hypertrophic scars, keloids and fibrotic diseases.

Many factors, such as oxygenation, infection, age, nutrition and diabetes, affect wound healing, leading to improper or impaired tissue repair. The contamination and/or colonization of wounds with bacteria such as Staphylococcus aureus (*S. aureus*), Pseudomonas aeruginosa (*P. aeruginosa*), and β-hemolytic streptococci, can cause prolonged wound inflammation, driving the wound to a chronic state. Often the bacteria in infected wounds occur in the form of biofilms, which are more resistant to conventional antibiotic treatment.

In other words, the proliferation of fibroblasts and differentiation of myofibroblast, and overcoming infection with antibiotic-resistant pathogens, are crucial in wound healing and wound closure. Impaired wound healing time is often correlated with chronic bacterial contamination in the wound area.

Because wound healing is tightly regulated by redox mechanisms, reactive oxygen and nitrogen species as the central components mediating biological effects of cold plasma are identified. A directed delivery of the plasma-generated species to cells and tissues works to commence new therapeutic options not only in the healing of pathological wounds but also related to other redox-based diseases.

Reactive species carried in CAP have antibacterial effects, which can act effectively against resistant bacteria in contaminated wounds. Moreover, ROS regulates intracellular oxidative stress, which promotes fibroblast differentiation processes; and nitric oxide, NO, regulates proliferation and differentiation in human skin cells.

Highly-reactive, short-living radicals, such as singlet oxygen, can generate a more stable reactive species such as H_2O_2 within the buffer after reacting with H_2O. Generated H_2O_2 positively affects proliferation, inhibition and reduction of myofibroblasts, and also provides antibacterial effects. It also induces differentiation in fibroblast by enhancing intracellular oxidative stress and/or causing cell cycle arrest.

In all, it is believed that the biologically-active reactive species carried in the CAP, as demonstrated by the APS treatment, can positively influence collagen synthesis and the activation of fibroblasts. During the inflammatory phase fibroblasts are involved in the secretion of cytokines and growth factors to activate the immune defense by increasing the accumulation of inflammatory macrophages and neutrophils. During the proliferative and remodeling phase fibroblasts are important for tissue granulation and reorganization of the provisional extracellular matrix (ECM). The exogenic

NO delivered by the CAP may function, similar to the endogenous NO release, to regulate collagen formation and wound contraction during the wound healing. For chronic wounds, the induced inhibition of cell proliferation may have beneficial therapeutic effects on the hyper-proliferative skin diseases, such as excessive scarring or psoriasis.

8.6 Chronic Wounds

Some wounds progress smoothly and rapidly through the phases of healing, while others stall, most often in the inflammatory phase, to become chronic. Factors that predispose to a chronic wound include (1) formation of biofilm adhering to the wound to protect the bacteria contained in it; (2) an imbalance in cytokines, growth factors and/or Matrix Metalloproteinases (MMPs), caused by abnormal increase of inflammatory mediators due to biofilm burden; growth factors and the extracellular matrix are necessary for healing to take place; (3) hypoxia, impeding fibroblast proliferation and collagen production which are crucial to wound healing, and providing environment for bacteria to flourish; (4) a catabolic state due to poor nutrition, causing decreased fibroblast formation.

APS is able to kill bacteria growing in biofilms in wounds to avoid becoming chronic wounds. Many chronic wounds represent a significant burden to patients, health care professionals, and the cost of the health care system. Common serious chronic skin and soft tissue wounds include the diabetic foot ulcer, the pressure ulcer, and the venous stasis ulcer. Invasive systemic infection caused by chronic wounds is a serious complication that increases mortality risk. Chronic wounds have decreased levels of growth factors and bacteria in the wounds grow to form biofilms which cause infections more difficult to treat.

ROS are essential particularly at low levels for a wide range of innate immune functions, including antiviral, antibacterial, and anti-tumor responses. It can effactually combat two common bacteria, Pseudomonas aeruginosa and Staphylococcus aureus, which show up frequently in wound infections, but are resistant to antibiotics because they have a protective layer called a biofilm. APS treatment of chronic wounds has the potential to cure biofilm infections and to accelerate wound closure. It instigates oxidation process to cause enzymatic degradation of extracellular DNA, which can weaken the biofilm structure. The treatment is very well tolerated with no apparent side effects. Nitric oxide delivered by APS can trigger the dispersal of biofilms. APS advances the aggregation of platelets to elevate the growth factor level, starting the healing cascade. It provides a new approach for chronic wound healing.

Many common chronic conditions result in hypoxia. Hypoxia allows certain negative entities, such as bacteria, to flourish and impedes fibroblast proliferation and collagen production, both of which are crucial to wound healing. APS provides OI in situ in the treatment, and so overcomes the hypoxia problem; in addition, also introduces a bactericidal environment.

The plasma treatment avoids the nasty side effects that drugs often bring; it destroys bacteria indiscriminately — whether it is antibiotic-resistant or not.

8.7 Burn Wounds

There are four different degrees of burn injury. First-degree burns appear red without blisters. When the injury extends into some of the underlying skin layer, it is a second-degree burn. Blisters are frequently present and very painful. In third-degree burn, the injury extends to all layers of the skin. Often there is no pain and the burn area is stiff. Healing typically does not occur on its own. A fourth-degree burn often appears black. It causes injury to deeper tissues, such as muscle, tendons, or bone, and frequently leads to loss of the burned part, which usually require surgical treatments, such as skin grafting.

In large burns (over 30% of the total body surface area), there is a significant inflammatory response. Such burns often require large amounts of intravenous fluid, due to capillary fluid leakage and tissue edema. Tetanus toxoid should be given to minimize the risk of infection complications. During the period of edema, the ability of the tissues to receive oxygen and nutrients is reduced, while susceptibility to infection is increased. Wound cleansing is given routinely.

APS is a gas state dry disinfectant, which is more comfortable than the solutions and also renders the treatment easy. APS also works to mitigate burn itch.

8.8 Wound Care

The main complication of an open wound is the risk of infection. An infected wound may show:

1) an increase in drainage;
2) thick green, yellow, or brown pus;
3) pus with a foul odor.

Other signs of infection include having:

1) a fever of over 100.4 °F for more than four hours;
2) a tender lump in your groin or armpit;
3) a wound that isn't healing.

The treatments include to drain or debride the wound and often to take an antibiotic if infection from bacteria develops. In serious cases, surgery may be required to remove infected tissue and sometimes the surrounding tissue as well.

Serious conditions that can develop from an open wound include:

1) Lockjaw; causing muscle contractions in the jaw and neck, is caused by an infection from the bacteria that cause tetanus.
2) A necrotizing soft tissue infection (gas gangrene); leading to tissue loss and sepsis (a systemic infection), is caused by a variety of bacteria including Clostridium and Streptococcus.
3) Cellulitis; an infection of the skin that is not in immediate contact with the wound.

Wound care is needed to prevent infection and facilitate wound healing. Understanding of the wound healing process and wound care is necessary for identifying abnormal healing processes and wound contamination, and providing proper treatment of common wounds. Good wound management includes assessing nutritional status, nutrient intake and appropriate intervention as needed.

If, in the hemostasis process, the wound is disinfected properly, wound healing in the inflammatory phase is shortened to minimize the risk of pus formation during healing. This is a salient feature of applying CAAP for bleeding control.

Factors promoting wound healing include:

1) adequate oxygenation;
2) adequate rest or local immobilization;
3) sufficient blood supply;
4) proper nutrition.

Problems

P8.1. Wound healing goes through four distinct but overlapping stages of the healing process, which are hemostasis, inflammation, proliferation, and remodeling. Many conditions can interfere the normal process to lead to poor wound healing. These include 1) venous insufficiency, diabetes, thrombocytopenia and other blood dyscrasias, which affect hemostasis, 2) an increase in exudate due to the presence of excess neutrophils, which prolongs the inflammatory phase, 3) dry and non-protective wound environment, which may result in prolongation of the proliferation phase, and 4) developing too slow epithelization of tissue over a denuded area; a scar will form.

Comment on how APS treatment improves conditions in each healing stage.

P8.2. Each phase of wound healing is characterized by the sequential elaboration of distinctive cytokines by specific cells. Give the distinctive indicator(s) of each phase.

P8.3. After the initial phase of hemostasis, the steps in the procession of wound healing include inflammation, the fibroblastic phase, scar maturation, and wound contracture. Wound contracture is a process that occurs throughout the healing process, commencing in the fibroblastic stage.

The inflammatory phase occurs immediately following the injury and lasts approximately 6 days. If the "cleansing" of the wound is incomplete, chronic inflammation can ensue, resulting in healing delay and prominent scarring. On the other hand, if the treatment of hemorrhage also provides disinfection, the inflammatory phase may be shortened to start the fibroblastic phase, the overall healing period can be reduced considerably and the risk of forming scar may be avoided. This was demonstrated by the APS plasma treatment for hemostasis; as shown in Fig. 8.3, plasma treated cross cuts heal much faster, and no apparent scars can be seen. Explain in detail the plasma effect on speeding up the healing process.

P8.4. It was understood that wound healing time could be decreased up to 50% if appropriate dose-dependent settings are created. Hyperbaric oxygen has been used to promote healing. Agents such as platelet-rich plasma (PRP) and erythropoietin (EPO) are modulators that have a positive effect on tissue regeneration and have been used successfully to enhance the healing of wounds. New techniques which can enhance the proliferation and migration of cells leading to the acceleration of the healing of wounds are being explored; laser and CAPs are examples. Analyze the common features of the APS technique and the hyperbaric oxygen and PRP techniques.

P8.5. Insufficient healing can result in a hypotrophic or atrophic scar formation. All wounds in adult skin heal with a scar. The degree of inflammation has a direct impact on the ultimate scar formation. Medications with the greatest risk of negatively affecting wound healing and skin integrity include antibiotics, anticonvulsants, angiogenesis inhibitors, steroids, and nonsteroidal anti-inflammatory drugs. Drugs that may aid wound healing, on the other hand, include ferrous sulfate, insulin, thyroid hormones, and vitamins.

Nutritional factors are also critical for proper wound healing. Improvement in the nutritional status of adults correlates with enhanced wound healing.

Make a list of likely negative impacts and positive impacts of those medications on wound healing.

P8.6. The purpose of analgesia is not simply to make patients feel better, but to facilitate early ambulation, adequate oxygenation and nutrition and in doing so to reduce the stress response to the trauma, encourage wound healing, and minimize the risk of developing chronic wound pain. Therefore, early stage pain relief works to optimal wound healing. CAAP treatment disinfects the wound to provide preemptive analgesia, which dampens the body's natural response to injury and thus decreases the body sensitivity to pain that follows. Outline assays to justify this possibility.

Chapter 9

Advanced Bleeding Control

9.1 Background

Blood vessels are the tubes in which arteries take blood under high pressure from the heart to the body and veins return blood under low pressure back to the heart via the lungs.

Traumatic injuries that damage an artery or vein can be very serious and result in life threatening bleeding. In particular, artery bleeding appears in spurts, rather than in a steady flow. The amount of blood loss can be copious, and can occur very rapidly.

Blood loss to become critical varies with individual; usually, it is considered serious for adults to lose 1000 cc, for children to lose 500 cc, and for infants to lose 250 cc. The first priority in the treatment of blood vessel wounds is hemostasis; according to the Department of Homeland Security, it can take less than five minutes for someone to bleed to death. It is also noted that hemorrhage is the leading cause of preventable death on the battlefield. Uncontrolled hemorrhage is also the second leading cause of death in civilian trauma patients. The search for efficacious treatment remains one of the most challenging problems facing emergency medical professionals. New first aid products designed to rapidly promote coagulation and arrest ongoing hemorrhage are needed to save preventable deaths.

The treatment must be thorough and absolute; otherwise, the danger of wound infection and subsequent grave danger to the patient's life could be inevitable. Treatment is dependent on the location of the injury. Larger arteries and arteries feeding important areas need emergent repair to prevent death of the tissue the artery was supplying with blood. In general veins are not sewn back together unless they are large.

It is universally accepted that hemostatic agents are the primary tool for smaller bleeding injuries; however, in the case of a serious injury resulting in extreme blood loss, a hemostatic agent alone would not be very effective. In hospitals or at facilities where proper medical attention is available, some main types of hemostasis used in emergency medicine include:

1. **Physical agents (gelatin sponge)** — These gelatin sponges absorb blood, allow for coagulation to occur faster, and give off chemical responses that decrease the time it takes for the hemostasis pathway to start. It can quickly stops or reduces the amount of bleeding. These physical agents are mostly used in surgical settings as well as after surgery treatments.

2. **Chemical/topical** — Microfibrillar collagen is the most popular topical agent used in surgery settings to stop bleeding. It attracts the patient's natural platelets and starts the blood clotting process when it comes in contact with the platelets. This topical agent requires the normal hemostatic pathway to be properly functional.

3. **Sutures and ties** — Sutures are used to close an open wound, keeping the injured area from pathogens and other unwanted debris from entering the site; it is also essential to the process of hemostasis. Sutures and ties join back skin, which reduces the surface area of the wound. It allows for platelets to start the process of hemostasis at a quicker pace and shortens the wound recovery period.

However, at the places where proper medical attention is not available immediately, for instance, on battlefields and at accident sites, the main types of hemostasis are

1. **Direct pressure** — This approach can slow down bleeding of artery injury, allowing for more time to get to an emergency medical setting. Soldiers use this skill during combat.

   Treatment consists of the application of local pressure to the wound by means of fingers to slow down bleeding. It is also recommended to elevate the wound part to reduce blood flow into that area. In the cases of brachial in arm and femoral in legs, one should use a finger to press down at a point on the major artery of the wound area to reduce blood flow.

2. **Pressure dressing** — Treatment consists of the application of local pressure to the wound by means of a surgically sterile dressing and this usually suffices to stop the bleeding. In case of damage to large arteries in limbs, it is best to apply some form of tourniquet to the limb immediately above the wound; however, the tourniquet can cause gangrene beyond the site of the wound if it is kept tightened too long.

Some protocols call for the use of clotting accelerating agents, which can be either externally applied as a powder or gel, or pre-dosed in a dressing, or as an intravenous injection. These may be particularly useful in situations where the wound is not clotting due to the size of wound or due to hemophilia.

9.2 Hemostasis in Combat Casualties

Methods of treating external hemorrhages in combat casualties depend on the locations and types of the injuries. Commonly, hemostatic gauzes, such as a combat gauze, which must be packed into wounds and compressed for several minutes, are applied directly to compressible wounds; the clotting rate and clot strength of a hemostatic gauze depend on the hemostatic agent and adopted dressing, which can be improved by the raw material, kaolin coating, and the size of absorbent sponge. Hemostasis normally begins with platelet aggregation and formation of a platelet plug, providing a foundation for a thrombus to then form after initiation of the coagulation cascade. The efficacy and acute safety of the dressings drop with severity of the wound. When hemostatic gauze is not able to withstand increasing arterial pressure, bleeding will continue under the dressing to form hematoma, which prevents blood clotting. PolySTAT, a synthetic polymer that enhances coagulation by cross-linking fibrin, can be impregnated into hemostatic gauze; it adds a fibrin cross-linking hemostatic mechanism to withstand increasing arterial pressure and increases blood absorption to improve blood clotting. However, the debris of PolySTAT is difficult to remove thoroughly, causing infection in the wound healing period. Usually, hemostatic gauze alone has limited or no effect for non-compressible wounds in junctional anatomic locations.

The Combat Ready Clamp (CRoC) has been developed and deployed for treating junctional hemorrhage on the battlefield. CRoC stops the hemorrhage and prevents re-bleeding with the aid of hemostatic clamp. However, removal of the clamp will promptly lead to re-bleeding, reducing the survival rate of the person being treated. In other words, CRoC is an effective hemostatic adjunct for compression and control of groin hemorrhage, but it cannot downgrade clamp-necessary wounds; a hemostatic clamp has to be used together with hemostatic gauze for junctional hemorrhage control. This hemostatic intervention seals the wound at the skin, allowing internal clot formation. However, applying a hemostatic clamp continuously for an extended period may cause minor inflammation on blood vessels (endothelium) and nerves.

In very severe situations, such as amputation of limbs, which involve high-pressure arterial bleeds, it is best to apply some form of tourniquet to the limb immediately above the wound if it is applicable. However, care must be taken that this is only applied just so tightly enough to arrest the bleeding, and that the tourniquet is loosened every 20 minutes to reestablish arterial flow. It reestablishes the peripheral circulation through collateral vessels to avert gangrene beyond the wound site; this is especially the case in the lower limb.

In any case, swift hemostasis is the key to improve the survival rate in combat injury. It also reduces blood loss, making wound healing easier with better results.

Therefore, it is vital that there be available a portable blood coagulator, for hemostasis on the injured soldier on the battlefield, as well as on the injured person in an accident. It is also desirable that the treatment improves the timing of the healing process so that healing will result in regeneration, rather than forming a scar.

9.3 APS as an Advanced First Aid Tool

The APS shown in Fig. 2.6a was redesigned as shown in Fig. 9.1a. The air blower was combined with the discharge module, so that the tubular air duct was eliminated, and the APS could easily unplug

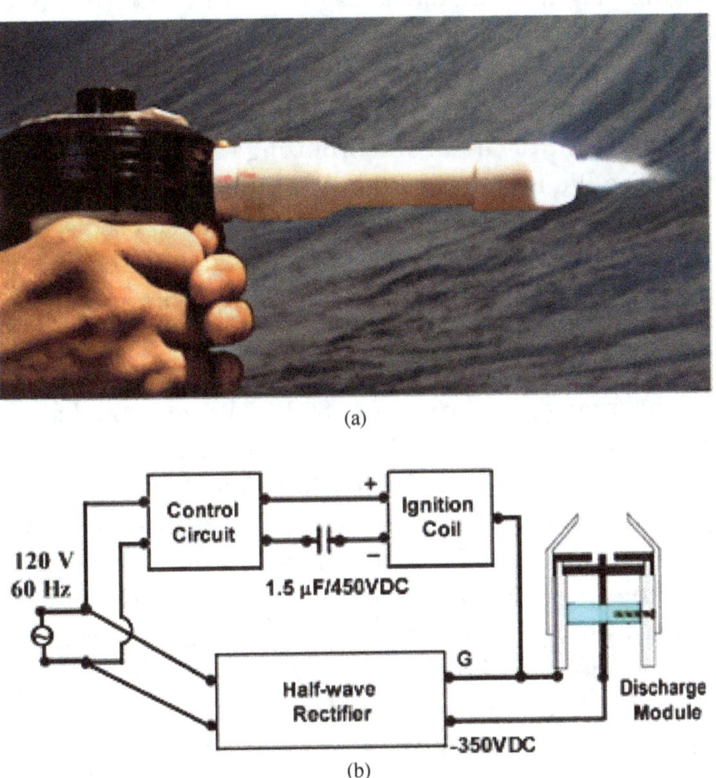

(a)

(b)

Fig. 9.1. (a) A portable air plasma spray and (b) a block diagram of the power supply.

from the power supply, making the setup easy to carry. The electric circuit shown in Fig. 2.6a is simple, but the transformer is not efficient and is heavy. The power supply also overruns the discharges to elevate the plasma temperature. As shown in Fig. 2.6b, the voltage to maintain the discharge in the 1mm gap between the electrodes, after air breakdown occurs (breakdown field ~3kV/mm), is less than 350 V. Thus the discharge module can be run with a pulsed dc power supply together with an ignition coil, which triggers air breakdown to set the subsequent low voltage pulsed discharge. But the low voltage pulse has to synchronize with the trigger pulse, and preferable starts right after the trigger pulse. This guides the design of a new power supply being used to run the new air plasma spray.

Shown in Fig. 9.1b is a block diagram of this new type power supply running APS efficiently. The 120 VAC input is half-wave-rectified to −350 VDC pulsing at the input frequency of 60 Hz (in the United States). A high voltage (HV) induction (ignition) coil is used as a trigger, and a control circuit synchronizing with the input frequency is used to fire the ignition coil whose output terminal is connected to the floating ground of the power supply. The 60 Hz discharge has a smaller duty cycle of about 8%, but much larger discharge current of about 12.5 A, compared with the corresponding values of 25% and 2.6 A of the discharge run by the power supply shown in Fig. 2.6a. The peak power in the discharge of the module is about 4 kW and the average power is about 80 W. The generated air plasma spray has a low temperature which is less than 320 K, yet it delivers an abundance of atomic oxygen to a large region.

A system similar to that shown in Fig. 3.1b is used to measure the intensity distribution of 777.4 nm emissions of the plasma plume. However, the system is upgraded by using a 2-nm passband interference filter with a center wavelength of 777.4 nm (from using a 10-nm passband interference filter with a center wavelength of 780 nm). The imaging process started with the cover completely lowered, and the bright emissions from the direct flames required the camera to be operated at its lowest sensitivity, with the aperture stopped to f/16 and the exposure time set to 0.1 s. Even at this shortest exposure time of 0.1 s, the image was an average over six discharges, which evened out the instantaneous fluctuation of the emission in the image due to the 60-Hz periodic operation.

The cover was then moved up by 2 to 3 mm (reduced from 6.4 mm) sequentially in each image, with the aperture and the exposure time adjusted accordingly at each stage. Thus the spatial resolution of the measurement was improved significantly. It proceeded until the cover blocked most of the flame region to detect the upward boundary of the emission using the camera at high sensitivity, with the aperture opened to f/1.2 and the exposure time set at 1 s.

The resulting images were then calibrated and combined into a composite image giving contours of the intensity distribution, ranging from $<10^4$ to $>10^9$ Rayleighs; the line plots in white of the intensity at various distances from the nozzle are superimposed in the figure presented in the middle of Fig. 9.2. An image of the plasma plume recorded by a video camera is included on the

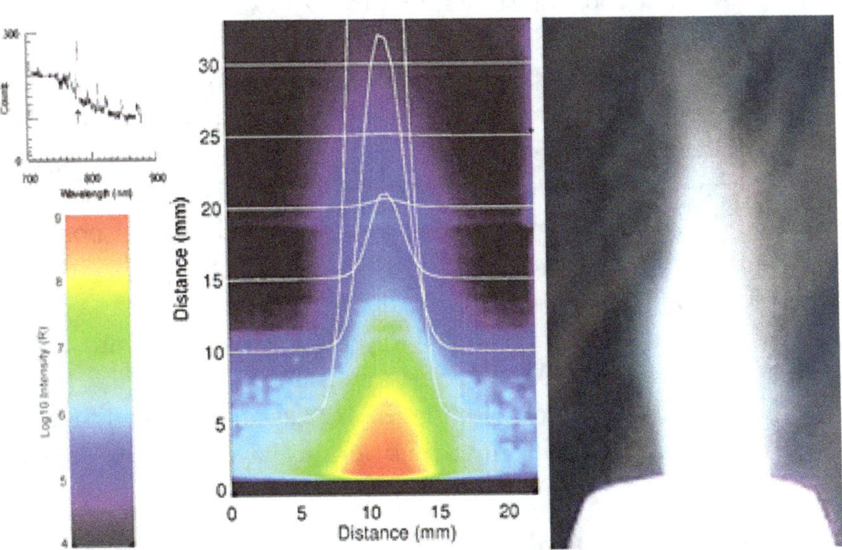

Fig. 9.2. Composite images showing the plume of the APS (on the right) comparing with the false-color calibrated images of 777.4-nm emissions from the APS (on the left) recorded by a narrow-band-filtered CCD camera, in which the line plots in white are the emission intensities at various distances from the nozzle. The insert at the upper left corner is the emission spectrum showing a peak at 777.4 nm.

right-hand side in Fig. 9.2 for reference. As shown, the visible plasma plume extends out axially to about 25 mm from the nozzle of the plasma generator, where the intensity of 777.4-nm emissions is about 10^5 Rayleighs, i.e., the apparent photon emission from a slice of the plasma plume at 25 mm away from the nozzle is about 10^{15} m^{-2} s^{-1}. The near-Gaussian appearance of the curves indicates that the radial structure in three dimensions is also near-Gaussian; it is a shape which is invariant under the Abel inversion and suggests being optically thick.

The discharge module can be redesigned to keep the required maintaining voltage of the discharge as low as 100 V. In this design, the input power source can be a 12V battery, so that the running of the device does not rely on the presence of an AC power outlet.

A DC pulse generator is then designed to run this air plasma spray. An inverter was used to transform 12V DC input to 100V DC pulses, which simultaneously generate synchronized 23 kV DC trigger pulses; each 100V pulse is added to a 23 kV trigger pulse as the output of the pulse generator, where the trigger pulse initiates air breakdown and the 100V pulse incites the subsequent electric discharge to maintain plasma generation. With the aid of a repetitive high voltage trigger operation, a matching circuit can be designed to match the dynamic plasma load to the source (i.e., the pulse generator). This battery-powered power supply enables this portable plasma device to operate anywhere. The average power is reduced to about 50 W and the generated air plasma spray also has a lower temperature which is less than 305 K, yet it delivers an abundance of atomic oxygen over a 30 mm exposure distance.

Pushing forward to developing the APS as an advanced first aid tool, in vivo trials of different serious wounds have been performed.

9.4 Animal Model Trials

In the following, tests and the results are presented. Adult pigs with blood pressure compatible to that of an adult human being, were used as animal models. Different types of large-scale potentially life-threatening wounds were treated by the hand-held air plasma spray

(APS) shown in Fig. 9.1a. Similar experimental arrangement as that described in Sec. 7.5, each pig was first injected with calmative-Stresnil and fastened on a table. The pig was then anesthetized with Isoflurance-Fluothane which kept it in a narcotized state, which was monitored throughout the entire trial period.

9.4.1 *Trial 1 — a deep cut at back*

Shown in Fig. 9.3a is a large and deep cut wound on the back of a pig. It involved a cut artery which causes main bleeding. Because a tourniquet couldn't be applied, one had to apply pressure above the

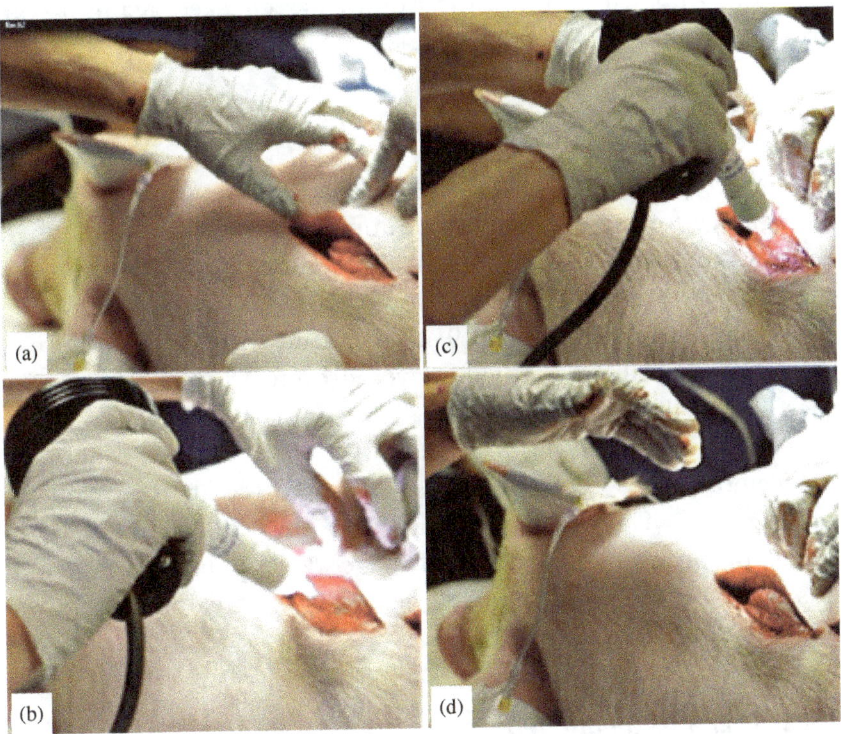

Fig. 9.3. APS treatment on a cut wound on the back; (a) examining the wound, and pressing on the cut artery to slow down bleeding, (b) treatment focused on the cut artery first, (c) moving APS to treat other an area of the cut, and (d) assessing the wound to confirm complete stop of bleeding.

wound to slow down the active hemorrhage. Thus, the first step was clearing away the blood with a medical cloth to locate the cut artery, and then using a finger to press the artery above the cut location as indicated in Fig. 9.3a.

APS was applied to first treat this cut artery, as shown in Fig. 9.3b, until its broken end was constricted and sealed to stop bleeding; subsequently, the entire wound area was treated by sweeping APS back and forth over the cut. The treatment continued as shown in Fig. 9.3c for a total time of about 25 s. The wound was then examined to see if hemorrhage was controlled.

As shown in Fig. 9.3d, all hemorrhage was stopped. Because the cut artery has a relatively small size, it took a very short time (~ 25 s) to seal it. After 30 minutes, the wound was checked again; it confirmed that there was no re-bleed.

9.4.2 *Trial 2 — a curved large cut in hind quarters area*

This large opening cut, shown in Fig. 9.4a, caused two broken arteries. Again, a tourniquet couldn't be applied, so pressure was applied above wound to slow down the active hemorrhage. Moreover, the cut arteries were hidden deeply inside the wound, which had to be opened up, as shown in Fig. 9.4a, to locate the arteries for treatment. The handheld APS was moved around to treat two arteries back and forth.

After about one minute treatment, the wound was examined, but hemorrhage was not yet controlled. APS was then applied for another 45 s. The second assessment of the wound, as shown in Fig. 9.4b, confirmed that all hemorrhage was stopped completely. In this case, it involved two cut arteries; thus the treatment time was increased to 105 s.

9.4.3 *Trial 3 — amputated leg*

The wound to be treated, as shown in Fig. 9.5a, was a complete amputation of a front leg at the middle joint. This is a life-threatening hemorrhage situation, it involves three broken major arteries.

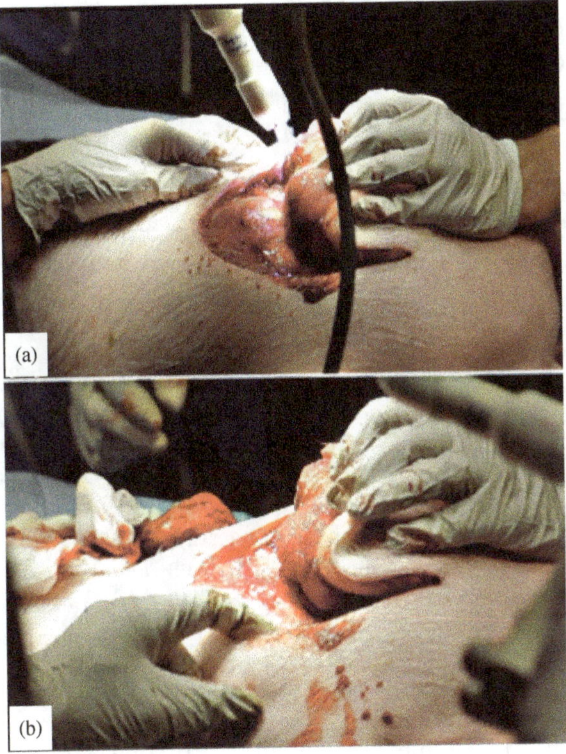

Fig. 9.4. APS treatment on a curved large cut wound in the hind quarters area; (a) treating the cut arteries back and forth continuously for about 105 s, and (b) assessing the wound to confirm no more bleeding.

Tourniquet was applied in advance before performing the operation to stop spurts of blood from the cut arteries, then APS was applied continuously to treat the cut arteries as shown in Fig. 9.5b. In this location, the arteries are protected between bones, which makes difficult to constrict the cut arteries. It took more than 5 minutes of continuous treatment, moving APS back and forth on three cut arteries, to stop all hemorrhage.

After the treatment, the tourniquet was loosened slowly to make sure that the wound did not re-bleed; the tourniquet was then removed as shown in Fig. 9.5c. This is a demonstration of successful downgrade of a tourniquet-necessary wound. It also meets the call for

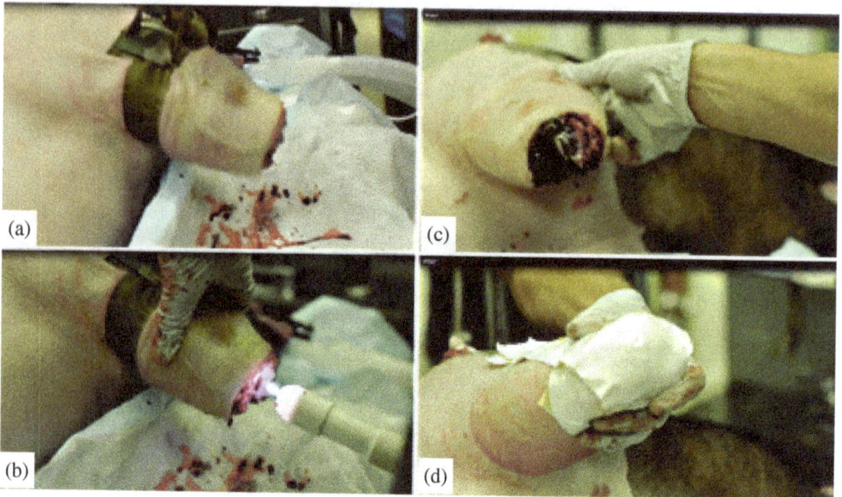

Fig. 9.5. APS treatment on an amputated leg; (a) tourniquet is applied to slow-down hemorrhage from an amputated leg, (b) APS is applied to the cut arteries back and forth continuously for more than 5 minutes, (c) after all hemorrhage is controlled, tourniquet can be removed, and (d) wound is wrapped by bandage before giving a formal treatment in the hospital.

"platinum ten", i.e., achieving bleeding control within 10 minutes, to ensuring the survival of wounded military personnel. The wound can be simply wrapped with bandage as shown in Fig. 9.5d for protection and preventing infection, before having a formal treatment in the hospital, as a protocol.

In the deep cut cases of trials 1 and 2, there were no apparent red clots were observed after the bleeding was controlled. It suggests that the hemostasis is mainly through vasoconstriction accelerated by the ROS. When atomic oxygen interacts with H_2O, reactive oxygen species are produced to induce a procoagulant state in endothelial cells by altering tissue factor (TF) structure. ROS suppress NO's vasodilator activity to accelerate vasoconstriction and degrades plasmin in anticoagulation to enhance platelet aggregation, which then contract tissue to seal injured blood vessels.

The capacity of APS in swift hemostasis and in downgrading hemostatic clamp/tourniquet necessary wounds can be further evaluated by battlefield simulation trials which are proposed in the following.

9.5 Proposed Battlefield Simulation Trials

Wounds caused by gun shot and blast on the battlefield are serious and life threatening. However, in most cases, if the hemorrhage is controlled in time, the lives of injured persons might be saved. Thus, using pigs as animal models to explore CAP efficacy on hemostasis of severe injuries encountered on battlefield will be of significance. Several trials simulating battlefield situations are proposed. Although not all of the CAP devices can control severe bleeding from the considered injuries for the trials; however, the results presented in Sec. 9.3 indicate that APS is a feasible device. Thus the proposed trials are based on the capacity of the APS. It is noted that the trials have to meet the ethic requirements and keep models in a narcotized state in all time.

9.5.1 *Proposed Trial 1 — close range shotgun wound to rear leg*

This can cause near complete amputation of a rear leg just above the middle joint, which imitates a blast like injury. The proposed treatment procedures are:

1. A tourniquet is applied immediately to stop high-pressure arterial bleeds.
2. CAP is applied continuously and swept back and forth among the broken arteries at a proper exposure distance (e.g., ~25 mm, but it depends on the device) to treat them all together.
3. It is better to constrict the broken arteries simultaneously, rather than one by one in sequence.
4. After a few minute treatment, the tourniquet is loosened slowly to examine whether the broken arteries are sealed.
5. If bleeding persists, one has to tighten the tourniquet to continue the treatment.
6. It is to repeat the process until all hemorrhage is controlled and the tourniquet can be removed.
7. Manipulating the limb to make sure that the broken arteries are sealed firmly, and the wound will not re-bleed.
8. Covering the wound with medical tape for post-test observation at later time, e.g., 1 hour later.

9.5.2 *Proposed Trial 2 — close range 7.62 mm gunshot through upper rump with large exit wound through groin*

This will cause significant internal active arterial hemorrhage; it cannot be stopped by applying a tourniquet, but it has to be stopped to avoid fatality. The proposed treatment procedures are:

1. The model is placed in a supine position.
2. Pressure is applied through the belly above the wound to slow down bleeding.
3. It is then inserting CAP into the exit wound, and applying it to the entire wound for, said 90 s, with continuous movement all around the wound.
4. After the treatment, pressure is released and the wound is examined to make sure all hemorrhage is controlled.
5. If the hemorrhage is not yet controlled, the steps from 2 to 4 are repeated; but the treatment time may be reduced.
6. After all hemorrhage is controlled, the wound should not be packed or covered immediately to allow for post-test observation.
7. After a giving, e.g., 1 hour, waiting time, the wound is reexamined to make sure no hemorrhage recurs at this wound site.
8. The wound is then packed with medical cloth for further post-test observation at the later time, e.g., 1 hour later.

9.5.3 *Proposed Trial 3 — dissection and laceration of femoral artery*

This type wound is located at a place where a tourniquet or a hemostatic gauze is not able properly applied. The proposed treatment procedures are:

1. After the cut, pressure is applied above the wound to slow down active hemorrhage.
2. A surgical clamp is applied at above the free end of the broken artery to stop the flow of blood.
3. CAP is then applied for, e.g., 30 s, to the injured blood vessel.

4. After the injured blood vessel is constricted visually, the surgical clamp is removed to check if the hemorrhage is completely stopped.
5. The wound is examined to make sure that there is no re-bleed with movement of the artery; reexamining it at a later time, e.g., 1 hour later.

9.5.4 *Proposed Trial 4 — dissection and complete severing of brachial artery*

Again, this type wound is located at a place where a tourniquet or a hemostatic gauze is not able properly applied. The proposed treatment procedures are the same as those listed for the proposed trial 3.

The results of successful trials can be applied for developing a CAP based first aid device and its operation protocol in treatment of different wounds. If the clinic trials are still necessary, these results will be useful reference.

9.6 Discussion

Because some injuries can cause a trauma patient to deteriorate rapidly, the time delay between injury and treatment has to be kept to a minimum. It is well established that the patient's chances of survival are greatest if he/she receives definitive care within a short period of time after a severe injury.

Historically, some four-fifths of military personnel killed in action perished within the first hour of injury, often because of blood loss. This hour — the golden hour — represents the span of time in which treatment of bleeding offers the greatest hope of survival. Therefore, bleeding control has been a battlefield medical treatment issue, which also calls for achieving control within as few as 10 minutes (so called platinum ten) to ensuring the survival of wounded military personnel.

First aid for critical traumatic patients in an emergency situation requires an effective medical tool, which is capable of stopping main hemorrhage of the wound within the initial 10 minute period, and to prevent infection. It is preferable that the first aid tool is portable, in particular, on the battlefield.

Moreover, the time that has elapsed since the wound occurred is an important factor in determining whether the wound should be closed primarily. Fast bleeding control together with proper disinfection in the treatment will allow some of wounds to be closed quickly without an increased risk of infection; that can reduce the patient's discomfort, speed wound healing, and decrease scarring.

The effectiveness of APS in rapid control of active life-threatening hemorrhage has been demonstrated in animal models, as presented in Sec. 9.4. The biocidal effects of APS also provide disinfection simultaneously in the treatment. The device can be made portable, using a battery as the power source. A device like this can meet the requirements of advanced first aid.

Further tests on wounds of artery cuts at different locations, as those proposed in Sec. 9.5 to simulate injuries occurred in the battlefield, will be useful to establish the treatment procedure for first aid in any critical situation.

9.7 Summary

Cold atmospheric plasma (CAP) is a new technology and an emerging research area for various medical purposes. Though full understanding of the chemical and physical mechanisms of CAP operation, and the biological pathways that CAP induces in cells and tissues during treatment, are subject for further studies, current R & D results with animal models and trends already show that CAPs have a large application spectrum including:

1. Prevention of healthcare-associated infections (HAIs) and hospital-acquired infections (HAIs);
2. Control of bleeding in emergencies, such as occurring on the battlefield and in an accident, and healing of chronic wounds;
3. Treatment of diseases in dermatology, dentistry, and oncology, and possibly enhancing drug uptake in chemotherapy; and
4. Cosmetology, etc.

Especially in first aid, a battery-powered portable APS device can be life-saving.

In medical applications, the risk assessment of the devices is essential. Consideration of risks includes

1. Thermal damage: CAPs are non-thermal, both the thermal and non-thermal components of the plasma effluent deliver thermal energy to the target; the limitation of the continuous treatment time should be imposed to avoid the possible impact of accumulated thermal energy on the target.

2. Radiation damage: UV-C poses the maximum risk, overexposure causes temporary skin redness and harsh eye irritation, but no permanent damage; its dose for healthy intact human skin should be below 3 mJ/cm^2. UV-B (in the range of 280-315 nm) may cause genetic mutations; prolonged exposure to UV-B can cause skin cancer, skin aging, and cataracts (clouding of the lens of the eye).

3. Biochemical damage: CAPs carry different concentrations of reactive species; some of those may be considered to be toxic for humans if they exceed a certain threshold, which is listed by The National Institute for Occupational Safety and Health (DC, USA). The most concerned toxic gases are

   (1) Ozone (O_3); the permitted exposure limit is 0.3 ppm for 15 min. and 0.1 ppm for 8 h continuous inhalation; and

   (2) NO_2 & NO; the limits are 5 ppm and 25 ppm, respectively, for 8 h continuous inhaling.

Problems

P9.1. Direct pressure, a technique to control bleeding from external bleeding sites, works even with bleeding from major vessels such as the carotid or femoral arteries. However, it must be applied consistently and with sufficient force to stop the bleeding.

If hemorrhage cannot be controlled with direct pressure, prompt application of a tourniquet should be performed. Tightening of the tourniquet should continue until arterial bleeding from the limb has stopped. The application and use of tourniquets to control bleeding before the onset of shock can result in lower mortality than that applied after the onset of hemodynamic instability. Properly applied tourniquets for a short period of time are safe; however, tourniquet application for longer periods of time may introduce potential adverse effects caused by complete limb ischemia.

What are the potential ischemic injuries resulting from prolonged tourniquet time periods?

P9.2. Under combat crossfire situations or during a mass casualty situation, what is the shortfall of applying topical hemostatic agents to control bleeding?

P9.3. Hemostatic agents work by causing vasoconstriction, promoting platelet aggregation, and contracting tissue, to seal injured blood vessels. Topical hemostatic agents are available in two forms — as a granular powder poured on wounds, or embedded in a dressing, which have been used in emergency bleeding control, especially in military medicine.

(1) Describe the working principles of Microfibrillar collagen hemostat (MCH) and Chitosan hemostat.

(2) What is the major difference of hemostatic pathways between the two?

(3) What are the functions of kaolin and zeolite used in hemostatic dressings?

P9.4. Bleeding refers to the loss of blood from blood vessels anywhere in the body. It is important to work quickly to control blood loss. Uncontrolled or severe bleeding can contribute to shock, circulatory disruption, or more serious health consequences such as damage to tissues and major organs, which can lead to death.

In the first aid, arterial bleeding usually requires finger-tip pressure at the point where the bleeding is coming from, rather than applying a generalized pressure on the wound itself for venous-type bleeding. In certain situations, a surgical clamp may be needed.

It is not generally recommended to use a tourniquet. However, in the case of severe injuries or severed limbs, a tourniquet may be needed to save a life.

(1) Explain why there are differences in treating venous-type bleeding and arterial bleeding.

(2) Explain why a tourniquet is not generally recommended if it is not necessary.

P9.5. Test results presented in Sec. 9.4 demonstrate the capacity for APS treatment to stop bleeding rapidly, and to downgrade tourniquet-necessary wounds. Summarize the likely activations and the benefits of the APS treatment.

Bibliography

Alexeff, I., and M. Laroussi, "The uniform steady-state atmospheric pressure dc plasma", *IEEE Transactions on Plasma Science*, **30**(1), 174–175, 2002.

Balzer J., K. Heuer, E. Demir, M. A. Hoffmanns, S. Baldus, P. C. Fuchs, P. Awakowicz, C. V. Suschek, and C. Opländer, "Non-thermal dielectric barrier discharge (DBD) effects on proliferation and differentiation of human fibroblasts are primary mediated by hydrogen peroxide", *PLoS ONE*, **10**(12), e0144968, 2015. https://doi.org/10.1371/journal.pone.0144968

Baxter, H. C., G. A. Campbell, A. G. Whittaker, A. Aitken, A. H. Simpson, M. Casey, A. C. Jones, L. Bountiff, L. Gibbard, and R. L. Baxter, "Elimination of TSE infectivity and decontamination of surgical instruments using RF gas-plasma treatment", *Journal of General Virology*, **86**(8), 2393–9, 2005.

Bekeschus, S., A. Schmidt, K. D. Weltmann, and T. Woedtke, "The plasma jet kINPen — A powerful tool for wound healing", *Clinical Plasma Medicine*, **4**(1), 19–28, 2016. https://doi.org/10.1016/j.cpme.2016.01.001

Chen, C. Y., H. W. Fan, S. P. Kuo, J. H. Chang, T. Pedersen, T. Mills, and C. C. Huang, "Blood clotting by low temperature air plasma", *IEEE Transactions on Plasma Science*, **37**(6), 993–999, 2009. doi: 10.1109/TPS.2009.2016344

Clark, R. A. F., *The Molecular and Cellular Biology of Wound Repair*, pp. 3–50, Plenum Press, New York, NY, 1996.

Costerton, J. W., P. S. Stewart, and E. P. Greenberg, "Bacterial biofilms: a common cause of persistent infections", *Science*, **284**, 1318, 1999. doi: 10.1126/science.284.5418.1318

Del Principe, D., A. Menichelli, W. De Matteis, S. Di Giulio, M. Giordani, I. Savini, and A. Finazzi-Agro', "Hydrogen peroxide is an intermediate in the platelet activation cascade triggered by collagen, but not by thrombin", *Thrombosis Research*, **62**(5), 365–375, 1991.

Dobrynin, D., G. Fridman, G. Friedman, and A. Fridman, "Physical and biological mechanisms of direct plasma interaction with living tissue", *New Journal of Physics*, **11**(11), 115020, 2009.

Duarte, S., S. P. Kuo, R. M. Murata, C. Y. Chen, D. Saxena, K. J. Huang, and S. Popovic, "Air plasma effect on dental disinfection", *Phys. Plasmas*, **18**(7), 073503 (7 pages), 2011. doi: 10.1063/1.3606486

Duske, K., I. Koban, E. Kindel, K. Schröder, B. Nebe, B. Holtfreter, L. Jablonowski, K. D. Weltmann, and T. Kocher, "Atmospheric plasma enhances wettability and cell spreading on dental implant metals", *Journal of Clinical Periodontology*, **39**(4), 400–407, 2012. doi: 10.1111/j.1600-051X.2012.01853

Fricke, K., I. Koban, H. Tresp, L. Jablonowski, K. Schröder, A. Kramer, K. D. Weltmann, T. von Woedtke, and T. Kocher, "Atmospheric pressure plasma: a high-performance tool for the efficient removal of biofilms", *PLoS ONE*, **7**(8), e42539, 2012. https://doi.org/10.1371/journal.pone.0042539

Fridman, A., A. Chirokov, and A. Gutsol, "Non-thermal atmospheric pressure discharges", *Journal of Physics D: Applied Physics*, **38**(2), R1, 2005.

Fridman, G., M. Peddinghaus, M. Balasubramanian, H. Ayan, A. Fridman, A. Gutsol, and Ari Brooks, "Blood coagulation and living tissue sterilization by floating-electrode dielectric barrier discharge in air", *Plasma Chemistry and Plasma Processing*, **26**(4), 425–442, 2006. doi: 10.1007/s11090-006-9024-4

Fridman, G., A. D. Brooks, M. Balasubramanian, A. Fridman, A. Gutsol, V. N. Vasilets, Halim Ayan, and G. Friedman, "Comparison of direct and indirect effects of non-thermal atmospheric-pressure plasma on bacteria", *Plasma Processes and Polymers*, **4**(4), 370–375, 2007. doi: 10.1002/ppap.200600217

Fridman, G., A. Shereshevsky, M. M. Jost, A. D. Brooks, A. Fridman, A. Gutsol, V. Vasilets, and G. Friedman, "Floating electrode dielectric barrier discharge plasma in air promoting apoptotic behavior in melanoma skin cancer cell lines", *Plasma Chemistry and Plasma Processing*, **27**(2), 163–176, 2007.

Fridman, G., G. Friedman, A. Gutsol, A. B. Shekhter, V. N. Vasilets, and A. Fridman, "Applied plasma medicine", *Plasma Processes and Polymers*, **5**, 503–533, 2008. https://doi.org/10.1002/ppap.200700154

Harman, D., "Aging: a theory based on free radical and radiation chemistry", *Journal of Gerontology*, **11**(3), 298–300, 1956.

Heinlin, J., G. Morfill, M. Landthaler, W. Stolz, G. Isbary, J. L. Zimmermann, T. Shimizu, and S. Karrer, "Plasma medicine: possible applications in dermatology", *Journal of German Society of Dermatology* (JDDG), Band **8**, 2010. doi: 10.1111/j.1610-0387.2010.07495.x

Herrmann, H. W., I. Henins, J. Park, and G. S. Selwyn, "Decontamination of chemical and biological warfare (CBW) agents using an atmospheric pressure plasma jet (APPJ)", *Physics of Plasmas*, **6**, 2284–2289, 1999.

Heslin, C., D. Boehm, V. Milosavljevic, M. Laycock, P. J. Cullen, and P. Bourke, "Quantitative assessment of bloodcoagulation by cold atmospheric plasma", *Plasma Medicine*, **4**(1–4), 153–163, 2014.

Hoffmann, C., C. Berganza, and J. Zhang, "Cold atmospheric plasma: methods of production and application in dentistry and oncology", *Medical Gas Research*, **3**:21, 2013. https://doi.org/10.1186/2045-9912-3-21

Iulianoa, L., A. R. Colavitaa, R. Leoa, D. Praticòb, F. Violia, "Oxygen free radicals and platelet activation", *Free Radical Biology and Medicine*, **22**(6), 999–1006, 1997. https://doi.org/10.1016/S0891-5849(96)00488-1

Kalghatgi, S. U., G. Fridman, M. Cooper, G. Nagaraj, M. Peddinghaus, M. Balasubramanian, V. N. Vasilets, A. F. Gutsol, A. Fridman, and G. Friedman, "Mechanism of blood coagulation by nonthermal atmospheric pressure dielectric barrier discharge plasma", *IEEE Transactions on Plasma Science*, **35**(5), 1559–1566, 2007.

Keidar, M., A. Shashurin, O. Volotskova, M. Ann Stepp, P. Srinivasan, A. Sandler, and B. Trink, "Cold atmospheric plasma in cancer therapy", *Physics of Plasmas*, **20**, 057101, 2013. https://doi.org/10.1063/1.4801516

Keidar, M., "Plasma for cancer treatment", *Plasma Sources Science and Technology*, **24**(3), 033001, 2015. https://doi.org/10.1088/0963-0252/24/3/033001

Kheirabadi, B. S., I. B. Terrazas, M. A. Hanson, J. F. Kragh, Jr., M. A. Dubick, and L. H. Blackbourne, "In vivo assessment of the Combat Ready Clamp to control junctional hemorrhage in swine", *Journal of Trauma and Acute Care Surgery*, **74**(5), 1260–1265, 2013.

Klämpfl, T. G., G. Isbary, T. Shimizu, Y. F. Li, J. L. Zimmermann, W. Stolz, J. Schlegel, G. E. Morfill, and H. U. Schmidt, "Cold atmospheric air plasma sterilization against spores and other microorganisms of clinical interest", *Applied and Environmental Microbiology*, **78**(15), 5077–5082, 2012. doi:10.1128/AEM.00583-12

Koinuma, H., H. Ohkubo, T. Hashimoto, K. Inomata, T. Shiraishi, A. Miyanaga, and S. Hayashi, "Development and application of a microbeam plasma generator", *Appl Phys Lett.*, **60**(7), 816–817, 1992; 10.1063/1.106527

Koritzer, J., V. Boxhammer, A. Schafer, T. Shimizu, T. G. Klampfl, Y. F. Li, C. Welz, S. Schwenk-Zieger, G. E. Morfill, J. L. Zimmermann, and J. Schlegel, "Restoration of sensitivity in chemo-resistant glioma cells by cold atmospheric plasma", *PLoS ONE*, **8**(5), e64498, 2013; doi: 10.1371/journal.pone.0064498

Kral, J. B., W. C. Schrottmaier, M. Salzmann, and A. Assinger, "Platelet interaction with innate immune cells", *Transfusion Medicine and Hemotherapy*, **43**, 78–88, 2016. doi: 10.1159/000444807

Kuo, S. P., D. Bivolaru, H. Lai, W. Lai, S. Popovic, and P. Kessaratikoon, "Characteristics of an arc-seeded microwave plasma torch", *IEEE Transactions on Plasma Science*, **32**(4), 1734–1741, 2004.

Kuo, S. P., *Portable Arc-seeded Microwave Plasma Torch*, United States Patent No.: **US 7091441 B1**, Aug. 15, 2006.

Kuo, S. P., O. Tarasenko, S. Nourkbash, A. Bakhtina, and K. Levon, "Plasma effects on bacterial spores in a wet environment", *New Journal of Physics*, **8**, 41(1–11), 2006.

Kuo, S. P., O. Tarasenko, S. Popovic, and K. Levon, "Killing of bacterial spores contained in a paper envelope by a microwave plasma torch", *IEEE Transactions on Plasma Science*, **34**(4), 1275–1280, 2006.

Kuo, S. P., S. Popovic, O. Tarasenko, M. Rubinraut, and M. Raskovic, "Fan-shaped microwave plasma for mail decontamination", *Plasma Sources Science and Technology*, **16**, 581–586, 2007. doi: 10.1088/0963-0252/16/3/017

Kuo, S. P., "Plasma assisted decontamination of bacterial spores", *The Open Biomedical Engineering Journal*, **2**, 36–42, 2008. doi: 10.2174/1874120700802010036

Kuo, S. P., O. Tarasenko, J. W. Chang, S. Popovic, C. Y. Chen, H. W. Fan, A. Scott, M. Lahiani, P. Alusta, J. D. Drake, and M. Nikolic "Contribution of a portable air plasma torch to rapid blood coagulation as a method of preventing bleeding", *New Journal of Physics*, **11**, 115016 (17pp), 2009. doi: 10.1088/1367-2630/11/11/115016

Kuo, S. P., *Portable Plasma Sterilizer*, US Patent No.: **US7777151 B2**, Aug. 17, 2010.

Kuo, S. P., C Y Chen, Chuan-Shun Lin, and Shu-Hsing Chiang, "Wound bleeding control by low temperature air plasma", *IEEE Transactions on Plasma Science*, **38**(8), 1908–1914, 2010.

Kuo, S. P., Todd Pedersen, and Travis Mills, "Two-dimensional distribution of atomic oxygen multiplet radiation produced by an air plasma torch", *IEEE Transactions on Plasma Science*, **39**(11), 2282–2283, 2011. doi: 10.1109/TPS.2011.2155089

Kuo, S. P., Cheng-Yen Chen, Chuan-Shun Lin, and Shu-Hsing Chiang, "Applications of air plasma for wound bleeding control and healing", *IEEE Transactions on Plasma Science*, **40**(4), 1117–1123, 2012. doi: 10.1109/TPS.2012.2184142

Kuo, S. P., "Air plasma for medical applications", *J. Biomedical Science and Engineering* (JBiSE), **5**, 481–495, 2012. Published Online September 2012. http://www.SciRP.org/journal/jbise/

Kuo, S. P., *Battery Powered Handheld Air Plasma Spray*, US Patent No.: **US8927896B2**, Jan. 6, 2015.

Kuo, S. P., "Air plasma spray for first aid", *Open Journal of Emergency Medicine*, **4**(3), 69–82, 2016. doi: 10.4236/ojem.2016.43010 and http://dx.doi.org/10.4236/ojem.2016.43010

Lai, W., H. Lai, S. P. Kuo, O. Tarasenko, and K. Levon, "Decontamination of biological warfare agents by a microwave plasma torch", *Physics of Plasmas*, **12**, 023501–6, 2005.

Laroussi, M., "Sterilization of contaminated matter with an atmospheric pressure plasma", *IEEE Transactions on Plasma Science*, **24**(3), 1188–1191, 1996. https://doi.org/10.1109/27.533129

Laroussi, M., I. Alexeff, J. P. Richardson, and F. F. Dyer, "The resistive barrier discharge", *IEEE Transactions on Plasma Science*, **30**(1), 158–159, 2002.

Laroussi, M., "Non-thermal decontamination of biological media by atmospheric pressure plasmas: review, analysis, and prospects", *IEEE Transactions on Plasma Science*, **30**(4), 1409–1415, 2002.

Laroussi, M., C. Tendero, X. Lu, S. Alla, and W. L. Hynes, "Inactivation of bacteria by the plasma pencil", *Plasma Processes and Polymers*, **3**, 470–473, 2006.

Laroussi, M., M. Kong, G. Morfill, and W. Stolz, edts., *Plasma Medicine: Applications of Low Temperature Gas Plasmas in Medicine and Biology*, Cambridge; New York: Cambridge University Press, 2012. ISBN 9781107006430 (hardback) and 1107006430 (hardback)

Laroussi, M., "From killing bacteria to destroying cancer cells: twenty years of plasma medicine", *Plasma Processes and Polymers*, **11**(12), 1138, 2014.

Laroussi, M., "Low temperature plasma jet for biomedical applications: a review", *IEEE Transactions on Plasma Science*, **43**(3), 703–711, 2015.

La Van, F. B., and T. K. Hunt, "Oxygen and wound healing", *Clinics in Plastic Surgery*, **17**(3), 463–72, 1990.

Leggett, M. J., G. McDonnell, S. P. Denyer, P. Setlow, and J. Y. Maillard, "Bacterial spore structures and their protective role in biocide resistance", *Journal of Applied Microbiology*, **113**, 485–498, 2012. doi: 10.1111/j. 1365-2672.2012.05336.x

Lu, X., G. V. Naidis, M. Laroussi, S. Reuter, D. B. Graves, and K. Ostrikov, "Reactive species in non-equilibrium atmospheric pressure plasma: generation, transport, and biological effects", *Physics Reports*, **630**, 1–84, 2016.

Machala, Z., K. Hensel, and Y. Akishev, eds., *Plasma for Bio-Decontamination, Medicine, and Food Security*, Springer The Netherlands 2012. ISBN 978-94-007-2909-4 (PB); ISBN 978-94-007-2851-6 (HB); ISBN 978-94-007-2852-3 (e-book); DOI 10.1007/978-94-007-2852-3.

Niemira, B. A., "Targeting biofilms with cold plasma: new approaches to a persistent problem", *Food Safety Magazine — BIOFILMS*, June/July 2017. https://www.Foodsafety magazine.com/magazine-archive1/junejuly-2017/ targeting-biofilms-with-cold-plasma-new-approaches-to-a-persistent- problem/

Sanaei, N., and H. Ayan, "Bactericidal efficacy of dielectric barrier discharge plasma on methicillin-resistant staphylococcus aureus and escherichia coli in planktonic phase and colonies in vitro", *Plasma Medicine*, **5**(1), 1–16, 2015.

Shintani, H., A. Sakudo, P. Burke, and G. Mcdonnell, "Gas plasma sterilization of microorganisms and mechanisms of action (Review)", *Experimental and Therapeutic Medicine*, **1**, 731–738, 2010.

Shohet, J. Leon, ed., *Encyclopedia of Plasma Technology*, 1st edition, Taylor & Francis Encyclopedia Program, CRC Press 2016.

Stoffels, E., A. J. Flikweert, W. W. Stoffels, and G. M. W. Kroesen, "Plasma needle: a non-destructive atmospheric plasma source for fine surface treatment of (bio) materials", *Plasma Sources Science and Technology*, **11**, 383, 2002. https://doi.org/10.1088/0963-0252/11/4/304

Stoffels, E., Y. Sakiyama, D. B. Graves, "Cold Atmospheric Plasma: Charged Species and Their Interactions with Cells and Tissues", *IEEE Transactions on Plasma Science*, **36**(4), 1441–1457, 2008; doi: 10.1109/ TPS.2008.2001084

Tarasenko, O., S. Nourkbash, S. P. Kuo, A. Bakhtina, P. Alusta, D. Kudasheva, M. Cowman, and K. Levon, "Scanning electron and atomic force microscopy to study plasma torch effects on *b. cereus* spores", *IEEE Transactions on Plasma Science*, **34**(4), 1281–1289, Aug. 2006.

USA Today, *Advanced First Aid For Troops Sought*, Monday, **A1**, September 14, 2009.

Weltmann, K. D., E. Kindel, R. Brandenburg, C. Meyer, R. Bussiahn, C. Wilke, and T. von Woedtke, "Atmospheric pressure plasma jet for medical therapy: plasma parameters and risk estimation", *Contributions to Plasma Physics*, **49**(9), 631–640, 2009. doi: 10.1002/ctpp.200910067

Weltmann, K. D., and T. von Woedtke, "Plasma medicine — current state of research and medical application", *Plasma Physics and Controlled Fusion*, **59**, 014031, 2017.

Woedtke, Th. Von, S. Reuter, K. Masur, and K. D. Weltmann, "Plasmas for medicine", *Physics Reports*, **530**(4), 291–320, 2013.

Woedtke, T. von, H. R. Metelmann, and K.-D. Weltmann, "Clinical plasma medicine: state and perspectives of in vivo application of cold atmospheric plasma", *Contributions to Plasma Physics*, **54**(2), 104–117, 2014. doi: 10.1002/ctpp.201310068

Index

advanced first aid, 13, 34, 116, 126, 129, 160, 168

air breakdown, 25, 26, 32, 34, 158, 160

air plasma, 4, 7, 11–14, 26, 36, 40, 41, 62, 65, 70, 79, 80, 87, 104, 107, 123, 124

air plasma spray (APS), 31, 34, 36, 47, 49–53, 55, 56, 64–67, 69–73, 75, 80, 116–119, 122, 123, 129–132, 136, 141, 144, 146–151, 157, 158, 160, 162–165, 168, 169

alkyl radical, 63

anthracis spores, 83, 86

anthrax, 11, 29, 82–84, 111

antibiotics, 9, 59, 63, 65, 70, 76, 147, 151

apoptosis, 54, 64, 65, 77, 139, 145

arc discharge, 4, 20, 27, 29, 31, 33, 36, 104, 105

arc seeded microwave plasma (ASMP), 31, 40, 87–89, 91, 93–96, 99, 103–106

atmospheric pressure plasma jet (APPJ), 22, 26, 36, 94, 95, 122

atomic force microscope (AFM), 97, 99, 100, 106

atomic oxygen, 9, 11, 12, 14, 30, 33, 38, 40–42, 44–47, 49, 50, 51, 53, 63, 70, 75, 76, 87, 103–107, 116, 117, 119–122, 129, 135, 136, 144, 145, 158, 160, 164

bacterial endospores, 39

bacterial spore, 8, 16, 60, 61, 83, 89, 90, 96, 97, 99, 104, 106, 107

biochemical reactions, 1

biofilm, 9, 10, 63, 65, 70–76, 80, 109, 146–148

Biological indicators (BIs), 17

biological warfare agents (BWA), 11, 82, 83

biologic indicator, 61

blood clot, 114, 116, 120, 124, 127

blood clotting index (BCI), 124, 125

blood coagulation, 128

breakdown threshold, 3, 4, 6, 16, 23, 25

cancer, 26, 42, 54, 64, 65, 76–78, 169

cascade impact ionization, 2, 3

chronic wound, 14, 76, 147, 148, 152, 168

coagulation, 12–14, 41, 64, 114–116, 119–124, 127, 130, 133, 135, 136, 144, 153, 154, 156

coagulation cascade, 12, 19, 40, 114, 115, 116, 121–123, 128, 129, 135, 136, 139, 155

cold atmospheric air plasma (CAAP), 13, 40, 41, 45, 76, 111, 124, 125, 137, 150, 152

cold atmospheric plasma (CAP), 7, 8, 19, 40, 47, 62, 65, 76–79, 83, 109, 110, 114, 120, 122, 123, 141, 146, 147, 151, 165–169

collagen, 54, 115, 127, 128, 135, 136, 139, 145–148, 154, 170

colony-forming unit (CFU), 64, 66, 68, 72–74, 92, 93, 96, 97

corona discharge, 3, 22, 25

cultivation, 68, 71

decontaminant, 11, 81, 87, 107, 111

decontamination, 7, 10, 11, 17, 18, 27, 29, 30, 39, 44, 62, 81–84, 87, 93, 94, 104, 107, 109–111

decontamination assurance level (DAL), 11

dental caries, 10, 60, 65

dental plaque, 65, 70

dielectric barrier, 6, 20, 21

dielectric barrier discharge (DBD), 20, 22–24, 64, 123

disinfectant, 9, 61–63, 65, 70, 78, 79, 84, 85, 149

disinfection, 7–9, 14, 16, 17, 27, 39, 56, 60, 61, 63–65, 71, 79, 144, 151, 168

DNA, 9, 38–40, 43, 56, 58, 63, 75, 77, 80, 85, 86, 103, 136, 148

DNA enzymes, 84

dormant, 70, 84

D value, 12, 17, 39, 40, 112

E. coli, 60, 64, 82, 84

electrical breakdown, 2–5, 108

electric discharge, 2, 19, 23, 44, 108, 160

electron-neutral collisions, 2

emission spectroscopy, 30, 33, 43, 45–47, 66, 87, 105

endospore, 60, 84, 85, 107

endothelial-derived relaxing factor (EDRF), 122

environmental scanning electron microscopy (ESEM), 72, 73

enzymes, 77, 110, 139, 144

erythropoietin (EPO), 151

fibrin, 40, 115, 120, 128, 135, 136, 139, 156

fibrin clot, 12, 128, 139

fibrinogen, 40, 113–115, 121, 127, 128, 135, 136

fibroblasts, 139, 140, 145, 146

floating-electrode dielectric barrier discharge (FE-DBD), 21, 63–65

fluorescence microscopy, 72, 74

fluorescent microscopy, 111

food-borne diseases, 109

free radical, 1, 19, 39, 42, 121, 139, 144

gangrene, 132, 149, 155, 156
germicidal effect, 10
gliding arc, 6, 7
glow discharge, 4, 25
gram-negative, 58–60, 79
gram positive, 58–60, 65, 70,
    79, 84

Hartley band, 46
healthcare-associated infections
    (HAIs), 8, 75, 168
hemorrhage, 12, 34, 114, 130, 131,
    151, 153, 155, 156, 162, 163,
    165–167, 168, 170
hemostasis, 13, 38, 53, 123,
    127–129, 131, 135, 136, 138, 140,
    150, 151, 153–157, 164
hemostasis plug, 40, 116
hemostatic, 166
hemostatic agent, 154, 155, 170
hemostatic clamp, 156, 164
hemostatic dressing, 170
hemostatic gauze, 155, 156, 167
hemostatic plug, 40
high level disinfection (HLD), 80
homeostasis, 13, 135
hospital-acquired infections
    (HAIs), 60, 75, 168
hydrogen peroxide, 38, 41–44, 61,
    63, 79, 83, 103, 106, 107, 121
hydrogen peroxide liquid
    (Vaprox), 112, 123
hydroxyl radical, 10, 38, 41–44, 63,
    103, 106
hyperbaric oxygen, 151
hyperbaric oxygen therapy
    (HBOT), 14
hypoxia, 144, 147, 148

inflammation, 140, 145, 150, 151,
    156
inflammatory, 13, 62, 138, 139,
    140, 141, 144, 146–148, 150,
    151

kill curve, 93–96
kINPen, 122

Lambert–Beer's law, 47

macrophages, 38, 139, 140, 146
metabolism, 13, 14, 77, 78, 135,
    136, 144, 145
microorganisms, 8, 10, 16–19, 39,
    53, 56, 60, 61, 63, 66, 68, 70, 72,
    73, 75, 76, 80–82, 84, 85, 90, 111,
    113
microwave discharge, 27
microwave plasma, 20, 29, 30
morphology, 72, 73, 84, 99, 103,
    106
myofibroblast, 140, 145, 146

neutrophils, 38, 113, 139, 140, 146,
    150
nitric oxide, 45, 46, 54, 67, 75, 76,
    135, 146, 148
nitrogen species (RNS), 7
non-equilibrium, 6, 7, 33, 38, 50,
    53, 62, 104

oral pathogens, 60, 65
overkill sterilization, 17, 61, 62
oxidant, 12, 54, 75, 77, 103, 115,
    119, 121, 122, 128, 135, 136
oxidation, 38, 40, 63, 75, 106–108,
    148

oxidation reduction process, 11, 107
oxidation, reduction (redox)
    reactions, 30, 39, 40, 87, 103
oxygen free radicals, 12, 120
ozone, 40, 46, 47, 54, 76, 104, 169

peroxyl radical, 42
Phenol coefficient, 78, 79
plasma, 1, 87
plasma bullets, 25
plasma jet, 5, 20, 22, 26, 39, 65
Plasma needle, 25
plasma pencil, 25, 39, 64
plasma spray, 5, 20, 33, 50, 71, 134
plasmin, 127, 135–136, 164
platelet clot, 122
platelet plug, 40, 115, 120, 127,
    128, 135, 155
platelet-rich plasma (PRP), 151
platelets, 12, 38, 40, 113–115,
    119–122, 127–129, 135–139, 40,
    148, 154
platinum ten, 164, 167
primary hemostasis, 115, 116, 120
proliferation, 13, 41, 136, 138, 140,
    146–148, 150, 151
protein degradation, 40, 85, 87,
    100, 103, 136

Rayleigh, 49, 50, 56, 159
reactive nitrogen species (RNS),
    19, 45
reactive oxygen, 63
reactive oxygen species, 10, 38, 42,
    63, 121, 139, 164
reactive oxygen species (ROS),
    7, 9, 11, 12, 14, 19, 38–42, 45,
    63–65, 76, 77, 79, 87, 104, 106,
    108, 120, 122, 135, 136, 144, 146,
    147, 164

reactive species, 7, 19, 22, 25, 26,
    30
redox, 146
regeneration, 7, 18, 42, 140, 144,
    151, 157
remodeling, 13, 138–140, 145, 146,
    150

scanning electron microscope
    (SEM), 85, 97, 106
secondary hemostasis, 115, 120
shock, 18, 60, 126, 137, 170
singlet oxygen, 38, 40, 45, 46, 104,
    106, 146
spectrometer, 45, 47, 51, 55, 122
spores, 39
sterile dressing, 132, 155
sterility assurance level (SAL), 8,
    17
sterilization, 7–9, 14, 16, 17, 30, 53,
    60–65, 76, 80, 83, 111
Sterilization Risk Assessment, 80
superoxide, 40, 42
superoxide radicals, 38
surface dielectric-barrier
    discharges (SDBDs), 20, 21, 25,
    36
surgical clamp, 166, 167, 171

thrombin, 12, 115, 120, 127, 128,
    135, 136, 139
Tissue factor pathway inhibitor
    (TFPI), 135
tissue factor (TF), 114, 135, 164
tourniquet, 13, 34, 132–134, 155,
    156, 161–167, 170, 171

ultraviolet, 85
UV-A, 54
UV-B, 54, 169

UV-C, 54, 169
UV radiation, 39, 47, 54, 63, 64, 85,
90, 105

vaporize, 112
Vaporized hydrogen peroxide
(VHP), 111, 112, 124–125
Vaprox, 124
V-I characteristic, 4

volume dielectric-barrier
discharges (VDBDs), 20
von Willebrand factor (vWF), 115,
128, 135

wound healing, 12–14, 34, 64, 129,
136, 138–140, 144–152, 156, 168

zone of inhibition, 68, 69, 71